About the Author

DON CORACE has been a successful real estate developer and businessman for more than twenty-five years. A sought-after public speaker, Corace has appeared on *Hannity & Colmes*, *The Neal Boortz Show*, and many other media venues, and has testified before Congress on property rights issues. He is the author of the novel *Offshore*, and he lives in Florida with his wife.

GOVERNMENT
PIRATES

Also by Don Corace

Offshore

GOVERNMENT
PIRATES

THE ASSAULT ON PRIVATE PROPERTY RIGHTS— AND HOW WE CAN FIGHT IT

DON CORACE

HARPER

NEW YORK · LONDON · TORONTO · SYDNEY

HARPER

GOVERNMENT PIRATES. Copyright © 2008 by NBE Entertainment, Inc. All rights reserved. Printed in the United States of America. No part of this book may be used or reproduced in any manner whatsoever without written permission except in the case of brief quotations embodied in critical articles and reviews. For information address HarperCollins Publishers, 10 East 53rd Street, New York, NY 10022.

HarperCollins books may be purchased for educational, business, or sales promotional use. For information please write: Special Markets Department, HarperCollins Publishers, 10 East 53rd Street, New York, NY 10022.

FIRST EDITION

Designed by Phil Mazzone

Library of Congress Cataloging-in-Publication Data is available upon request.

ISBN 978-0-06-166143-3

08 09 10 11 12 ID/RRD 10 9 8 7 6 5 4 3 2 1

To my wife, Ammi, for all her love and support . . .
and to our three children, Natalie, Brandon,
and Erik, the joy of my life.

CONTENTS

GOVERNMENT
PIRATES

INTRODUCTION: PROPERTY FOR THE TAKING

" . . . nor shall private property be taken for public use without just compensation."

— FIFTH AMENDMENT, U.S.
CONSTITUTION (TAKINGS CLAUSE)

June 23, 2005, was a very dark day in our nation's history. It was the day four men and one woman, dressed in black robes and sitting in a marbled temple in Washington, handed the government another weapon to continue its assault on our private property rights.

The U.S. Supreme Court ruled five to four in *Kelo v. New London, Connecticut* that the city could use its powers of eminent domain to promote economic development by "taking" waterfront homes and businesses and handing them over to a private developer to build a luxury hotel and upscale condos. A dangerous legal precedent had been set. The nation was outraged.

Newspaper editorials throughout the country attacked the decision:

- A *Richmond Times-Dispatch* (VA) headline read "Court-Endorsed Theft."

- A *St. Petersburg Times* (FL) editorial read "Eminent Mistake."
- A *Chattanooga Times Free Press* (TN) editorial wrote "Your Home, Freedom Attacked."
- And the *Hartford Courant* (CT) wrote "A Sad Day for Property Rights."

There were, however, a few exceptions. In a *New York Times* editorial titled "The Limits of Property Rights," the paper said the ruling was "a welcome vindication of cities' ability to act in the public interest" and a "setback to the 'property rights' movement, which is trying to block government from imposing reasonable zoning and environmental regulations." The *Washington Post* also sided with the *Times*.

On Fox TV's *Hannity & Colmes*, conservative Sean Hannity and liberal Alan Colmes vowed to expose eminent domain abuse. Rush Limbaugh, libertarian radio host Neal Boortz , and consumer advocate Ralph Nader were all critical of the decision. Even columnists like the late Molly Ivins, an activist for the left, and George Will, a staunch conservative, opposed the ruling.

The media attention even caused a backlash against two of the five U.S. Supreme Court justices who voted in favor of the decision. In New Hampshire, that state's Libertarian Party initiated a drive to use eminent domain to take Justice Stephen G. Breyer's 167-acre vacation homestead in Plainfield and turn it into a park. In Weare, a developer proposed to have the town turn Justice David Souter's home into the "Lost Liberty Hotel."

A poll conducted by NBC and the *Wall Street Journal* in July 2005 regarding Supreme Court issues revealed that Americans cared more about private property rights than any other issue—including the state right-to-die laws and parental notification for abortions. A number of Internet surveys by CNN, MSNBC, the *Christian Science Monitor*, and other major news organizations showed that more than 90 percent of those polled opposed the government seizure of private property to turn it over to developers.

Congress, of course, jumped on the bandwagon. A week after the decision, far-left California Democrat Representative Maxine Waters joined far-right Texas Republican Representative Tom Delay in supporting a *temporary* appropriations amendment which barred federal Community Block Grant funds for any city that did not prohibit eminent domain seizures for private development. A similar symbolic measure passed in the Senate.

Some states also took immediate action. Alabama and eight other states passed legislation to prohibit government from condemning property in nonblighted areas and transferring it to private developers.

Clearly, the *Kelo* ruling had hit a raw nerve.

Despite the widespread fury from conservatives, libertarians, and liberals alike, hundreds of cities throughout the country cheered the ruling and continued their assaults:

- The City of Riviera Beach, Florida, moved one step closer to displacing an estimated 6,000 local residents to build a billion-dollar waterfront yachting and housing complex.
- City officials in Freeport, Texas, began legal proceedings to seize two seafood companies to make way for an $8 million private boating marina.
- In Arnold, Missouri, the city initiated a plan to demolish thirty homes and fifteen small businesses to make way for a Lowe's home improvement store and a strip mall.
- The City of Oakland, California, evicted the owner of a family-run tire shop who refused to make way for a new housing development.

- Legal proceedings commenced against an apartment builder to take 154 acres of vacant land because the Town of Ridgefield, Connecticut, preferred corporate office space.
- City commissioners in Hollywood, Florida, seized a bank parking lot to make way for an exclusive condo tower.

Over the next several months, talk of eminent domain reform continued to sweep the nation. The U.S. House of Representatives passed a bill by a vote of 378 to 38 to restrict federal funding to cities that utilized eminent domain to benefit private developers. Thirty states passed some form of legislation attempting to counter the *Kelo* decision. Despite these efforts, by the end of 2006 the U.S. Senate had not even allowed a bill to reach the Senate floor for a vote.

Realistically, any new laws will not be foolproof. Enterprising lawyers will always find loopholes. The already overloaded court system will become even more clogged with thousands of cases. The nation's more activist judges, armed with the *Kelo* precedent, will chip away at any legislative measures to curb abuse.

Arrogant and corrupt city and county officials—with near limitless legal budgets—will continue to align themselves with well-heeled developers, political cronies, and major corporations to prey on the politically less powerful and disenfranchised, particularly minority communities.

This is not to say that eminent domain cannot be a useful tool for "public use" as stated in the Constitution's Fifth Amendment. Even many of the citizens who have had their properties unfairly seized and handed over to private developers agree with the long-held standard that eminent domain can, and should, be used for building roads, dams, airports, schools, military bases, and other necessary public uses—as long as owners are fairly compensated.

Kelo has sparked a healthy dialogue, but eminent domain abuse is only the "tip of the iceberg" when it comes to the assault on our property rights. Through local zoning and the regulation of wetlands and endangered species, governments take property without compensating owners and also *extort* land and money in return for approvals.

Broadly defined as "regulatory takings," these cases can be broken down into two categories: No-Compensation Takings and Pay-To-Play-Takings, or exactions. They involve complex land use and environmental laws that politicians and bureaucrats with no-growth agendas, Not In My Backyard (NIMBY) advocates, and environmental extremists can—acting alone or in concert—use to strangle property owners.

Here are some examples:

- A restaurant owner in Pompano Beach, Florida, endured thirty-one years of litigation because neighbors didn't want their ocean views blocked by a planned hotel.
- In Dartmouth, Massachusetts, a family's dairy farm was lost to foreclosure after being targeted by preservation groups.
- A 70-year-old man in Michigan was branded a "dangerous criminal" and faced prison and fines of $13 million for moving sand around his property.
- In the Florida panhandle, a father and son building a home spent 21 months in prison for cleaning out a ditch.
- A developer in Austin, Texas, was stopped from building a Wal-Mart after cave-dwelling spiders and beetles on the property were declared endangered species.
- In Colton, California, an endangered species of fly (yes, a fly!) halted the construction of a regional medical center.

In thousands of cases like these, property owners not only go through considerable emotional distress but are also forced to pay several thousands of dollars to litigate their claims in court. Why? Because politicians and bureaucrats—using *our* tax dollars—can. *afford* to involve owners in lengthy litigation and force property owners to bend to their will.

The federal court system has contributed to the problem and is in need of reform. For example, there are laws that require property owners who allege their constitutional rights have been violated to spend years in state courts before their claims can even be heard in federal courts. Porn peddlers in land use disputes have direct access to federal courts, but not other property owners!

Whether it is eminent domain, No-Compensation Takings, or Pay-To-Play Takings, the root of the problem is the nearly unbridled power of federal and Supreme Court judges. These judges have become the undisputed authority or "final arbiter" of interpreting the Constitution— although *nowhere* in the Constitution does it grant them these powers.

Several judges, who are appointed for life and are therefore unaccountable to the people, continue to broadly interpret the original intent of our Nation's founders and essentially dictate public policy. In turn, they open the door to power-hungry elected officials and bureaucrats who allow NIMBYs, environmental extremists, and other special interest groups to trample upon the Constitution.

We, the People, endowed by our Creator with certain unalienable rights, must protect our property from these government pirates!

1

THE ALMIGHTY JUDGES

*"It is emphatically the province and duty of the judicial
[branch] to say what the law is."*
—CHIEF JUSTICE JOHN MARSHALL (1803)

Alexander Hamilton wrote in the *Federalist Papers* that,
of the three branches of government, "the judiciary . . .
will always be the least dangerous to the political rights
of the Constitution" and the Supreme Court and federal
courts are "beyond comparison the weakest of the three
[branches] of power." Hamilton was dead wrong.

- Local government can seize private property through
  eminent domain and hand it over to a developer in
  order to generate more tax revenues.
- Property owners with local zoning disputes are forced
  to make their way through state courts before having
  cases heard in federal court—even though they allege
  their federal constitutional rights have been violated.
- Bureaucrats utilize their broad powers to regulate
  nearly every type of land holding water—from puddles,
  drainage ditches, ponds, creeks, and rivers—as a tool to
  set aside property from being developed without compensating owners.

• Landowners with an endangered species on their
  property not only receive no compensation for the
  loss of land to maintain a species habitat but in some
  instances must also pay to protect it.

Despite these and other reckless court decisions, former
Associate Supreme Court Justice Sandra Day O'Connor
has said, "we must recommit ourselves to maintaining the
independent judiciary that the Framers [of the Constitu-
tion] sought to establish." Indeed, our forefathers wanted
the judiciary to be independent of the executive and legislative branches of government. However, *nowhere* in the
Constitution is it written that a Supreme Court judge, who
is appointed for life, is to be the "ultimate arbiter" of interpreting the Constitution.

So how did the judges get this power? Through a legal
concept that all first-year law students learn—"judicial review."

JOHN MARSHALL'S SLEIGHT OF HAND

The Constitution established three branches of the federal government: the executive, legislative, and judiciary.
This was done to provide checks and balances because the
Framers understood, as Thomas Jefferson said, that when-
ever a man casts his eye on an office, "a rottenness begins
in his conduct."

Thus, the Constitution was designed so that the President and Congress could be held accountable to the people
through the ballot box—and to each other. Supreme Court
justices and federal judges, however, were to be independent of political motivations.

It is important to note that when the Framers met in
Philadelphia at the Constitutional Convention in 1787 and

took up the question of creating the federal judiciary, the delegates looked at the model established by the State of Virginia known as a "council of revision." Based upon the Virginia Plan, the Framers developed the concept that the federal judiciary could review and accept or reject acts of Congress, but the House of Representatives and the Senate could pass bills to *override* the decisions of the judges.

There was vigorous debate, but the delegates ultimately rejected the idea of the judiciary reviewing legislative acts. Therefore, *nowhere* in the Constitution does it specify that the judicial branch has the authority to override congressional acts. That all changed in 1803.

In 1801, during his last few hours in office, President John Adams had made a series of "midnight appointments" to fill as many government posts as possible with Federalists. One of these appointments was William Marbury as a federal justice of the peace. However, Thomas Jefferson became president before the appointment was officially given to Marbury.

Jefferson, a Republican, instructed Secretary of State James Madison not to deliver the appointment. Marbury sued Madison to get the appointment he felt he deserved and asked the court to issue a *writ of mandamus* requiring Madison to deliver it. The Judiciary Act, passed by Congress in 1789, permitted the Supreme Court of the United States to issue such a writ.

Since Marshall and the other Federalist justices wanted to expand the court's powers and they knew Jefferson did not want them to issue the writ, the high court voted unanimously not to issue it. Their basis: the Judiciary Act was unconstitutional.

Jefferson was pleased. However, what he did not understand at the time was that a dangerous legal precedent had been established: the Supreme Court, declaring an act unconstitutional, could now override Congress.

Legal scholars and students continue to debate *Marbury v. Madison* and the doctrine of judicial review. One thing is certain: Marshall took away one power of the court—to issue the writs—but gave it another far-reaching power. Now the Supreme Court would have the undisputed authority to have the final say of interpreting the Constitution—not elected officials accountable to the people.

The next step was for the high court to assert power over the states. Article 10 of the Constitution states: "The powers not delegated to the United States by the Constitution, nor prohibited by it to the States, are reserved to the States respectively, or to the people." In order to assert its control over the states, the Marshall Court pointed to a clause, called the "elastic clause," in Article 1 that said Congress could "make all laws which shall be necessary and proper."

This "implied powers" clause became the center of controversy that ignited the fierce debate between Federalists like Marshall and Hamilton ("the loose constructionists" who wanted to centralize power in the federal government) and Jefferson and the anti-Federalists (the "strict constructionists," who believed states should retain more rights).

During its thirty-four-year reign, the Marshall Court reviewed and ruled in several cases brought by states. This set another dangerous precedent: the federal judiciary could now review and accept or reject state laws—again, not elected officials accountable to the people.

Because of this unbridled power, some of today's federal judges and Supreme Court justices continue to broadly interpret the Constitution and, essentially, "legislate from the bench." As a result, we have decisions like *Kelo*. How do we reform this judicial tyranny, particularly as the Supreme Court—thanks to Marshall—can review and strike down acts of Congress it believes to be unconstitutional?

THE SUPER-LEGISLATURE

President Franklin Roosevelt was the last president who tried to reform the Supreme Court. In the 1930s, Roosevelt was attempting to pass his New Deal agenda to pull the country's economy out of the Depression but was being hindered by the Supreme Court. Since the Constitution was silent on the number of justices who can serve, Roosevelt figured Congress could make legislation to add (or "pack") the court with additional justices aligned with his philosophy. It created quite a stir.

Roosevelt used one of his famous fireside chats to sell his idea to the public. He lambasted the Supreme Court, calling it the "third house of Congress." He even borrowed a phrase from one of the high court justices that the court was a "super-legislature." Roosevelt warned, "We want a Supreme Court which will do justice under the Constitution, not over it. In our courts we want a government of law, not of men."

Roosevelt's broadcast created a furor. Even his vice president, John Nance Garner, worked actively against the measure to add justices, putting him at odds with the president and in league with leading liberal Democrats who were also in opposition. As result, both houses of Congress rejected the idea, and ultimately the proposed legislation was dropped. Although Roosevelt's prestige was seriously damaged, one of the four Supreme Court justices retired, and two others (who were considered swing votes) began to vote in favor of the New Deal agenda.

Contemporary constitutional lawyers, such as talk radio host Mark Levin, and scholars have proposed ways to rein in the power of the Supreme Court. Some reform suggestions are:

- to abolish lifetime appointments by instituting term limitation;

- to revise the confirmation process not only to include the investigation of judges being nominated but to review the conduct of sitting judges; and
- to give Congress the power to veto Supreme Court rulings.

In today's highly charged, circus-like partisan political environment, these and other such reform measures would be nearly impossible to accomplish. The only hope would be for the majority of the Supreme Court justices to agree to limit their judicial review authority.

What government body, however, is willing to *weaken* itself?

Part I

EMINENT DOMAIN

2
THE DESPOTIC POWER

"The right of property is the guardian of every other right, and to deprive a people of this, is in fact to deprive them of their liberty."

—DR. ARTHUR LEE, A VIRGINIAN AND DIPLOMAT DURING THE REVOLUTIONARY WAR

The power of eminent domain can be a valuable tool to benefit the public in the hands of responsible elected officials. In the wrong hands, however, it can serve as a wrecking ball to the American dream.

In 1795 the U.S. Supreme Court declared eminent domain a "despotic power." The justices understood that the power given to government to take private property could lead to abuse. They were right. Through legal precedent, the "public use" clause of the Fifth Amendment has been twisted to benefit private developers at the expense of individual private property rights.

THE EVOLUTION OF THE LAW

One of the most cited cases in eminent domain law is a 1954 case known as *Berman v. Parker.* Congress had passed the District of Columbia Redevelopment Act of 1945 which authorized the establishment of a redevelopment agency to eliminate slum and substandard housing conditions and promote commercial development.

Property owners in a predominantly black area in Washington, D.C., filed suit in circuit court to stop the agency from seizing their homes and businesses through condemnation due to blight. After losing in lower courts, they appealed to the U.S. Supreme Court.

The high court sided with the redevelopment agency and said that Congress or legislatures—and not the courts—were the best judges when it came to "social legislation." They also permitted some private development in redevelopment projects under the pretext that it was for a "public purpose." What was not spelled out, however, was a clear definition of what constituted "blight." The court concluded that this definition was for the legislatures to decide.

Another important case was a 1981 decision in the *Poletown Neighborhood Council v. City of Detroit* that involved General Motors. The auto maker announced its intention to close an obsolete plant, thereby eliminating more than 6,000 jobs. The company offered to build a new assembly plant in the city if a suitable site could be found. Detroit agreed and its redevelopment agency selected Poletown, a residential neighborhood, and began condemnation proceedings.

Residents filed suit. The case reached the Michigan Supreme Court, which ruled that transferring property to General Motors was a public use because of the economic benefits: jobs and increased tax revenue for the city.

Poletown had broad, national implications. It set the precedent that, for the most part, courts could not interfere with the government's determination of "public use." The decision ushered in an era that saw governments interpreting public use to fit redevelopment agendas.

The 1984 case known as *Hawaii Housing Authority v. Midkiff* involved neither blight nor creation of jobs. The Hawaii Legislature passed a land reform act to reduce what it perceived to be a land oligopoly—a concentration of ownership by a select few—that was traced to the early high chiefs of the Hawaiian Islands. The act created a condemnation scheme that transferred the title from these native Hawaiians to those people who had been leasing the lands.

The U.S. Supreme Court concluded that legislatures—both state and federal—should determine whether a taking would serve a public use. It declared that transferring ownership of the property was constitutional. Ironically, in the Hawaiian example, once the transfers were made, many of the new landowners reaped a windfall by selling their holdings to Japanese investors who built expensive vacation homes.

Like the *Poletown* case, the 2004 case of *County of Wayne v. Hathcock* was also a Detroit-area redevelopment case. It involved the condemnation of several properties to build a 1,300-acre business and technology park.

The affected property owners challenged the taking, saying that the project was not for a public purpose. The county argued that the precedent set by *Poletown* (twenty-three years earlier) gave it the legal right to use its powers of eminent domain to acquire the properties.

In a stunning reversal of *Poletown,* the Michigan Supreme Court ruled that the government's power of eminent domain must be in the interest of bona fide public use rather than the ill-defined notion of "public purpose" or "public benefit." It was a clear victory for property rights. Unfortunately, whatever legal protections the ruling provided were erased a year later by *Kelo v. New London, Connecticut.*

* * *

THE HUNGER FOR TAX REVENUES

In the words of the Connecticut Supreme Court, the City of New London approved a development plan that was "projected to create in excess of 1,000 jobs, to increase tax and other revenues, and to revitalize an economically distressed city, including its downtown and waterfront areas."

As in nearly all redevelopment projects, cities and counties form redevelopment agencies whose members are local elected officials. In the 1954 *Berman* case, for example, Washington, D.C., city council members sat on the redevelopment board. In the *Kelo* case, however, the City of New London created the New London Development Corporation (NLDC).

The key distinction between the two cases was that New London *delegated* its powers of eminent domain to the NLDC—a private, nonprofit corporation. To say that this maneuver could become a fertile ground for abuse would be an understatement.

Another difference between the cases was that *Berman* involved blighted properties and *Kelo* did not. In fact, the New London officials never claimed that the properties to be condemned were in poor condition or in any way jeopardized the health, welfare, and safety of the community. This was all about increasing tax revenues and creating jobs.

The U.S. Supreme Court sided with the city in a 5-to-4 decision. The majority of the justices found that the courts should not "second-guess the wisdom of the . . . city" and that private development *could* benefit the public under the pretext that it *might* generate more tax revenue and create more jobs.

Justice Clarence Thomas, known to be one of the more conservative justices on the court, feared the precedent *Kelo* would set. In his dissenting opinion, he wrote that it is wrong "when the government takes property and gives it to a private individual, and the public has no right to use the property."

Not unexpectedly, *Kelo* "opened the floodgates" for more abuse. Within a year, more than 5,700 properties nationwide had been threatened or taken by eminent domain for private development—compared to an estimated 10,000 examples over the five-year period prior to the ruling.

Eddie Perez, mayor of Hartford, Connecticut, argued that the *Kelo* ruling did not set a new precedent. Before a Senate committee, Perez, speaking on behalf of the National League of Cities, stated that the court "simply reaffirmed years of precedent that economic development is a 'public use' under the [Fifth Amendment]."

There is no question that revitalizing communities by preventing or eliminating slums, generating jobs, and increasing tax revenues are worthwhile goals. However, beyond the legal debate, more than 90 percent of Americans polled are dead set against the *Kelo* ruling. Coupled with the increasing distrust of its elected officials by the American public, the redevelopment industry has gotten a black eye—and for good reason.

3

REDEVELOPMENT: IS IT A SCAM?

"[Redevelopment is] like playing Monopoly with real money, it's like playing Legos with real bricks and mortar. And, on occasion, you can even help your son."
—LAS VEGAS MAYOR OSCAR GOODMAN

Cities and counties throughout the country hail the virtues of redevelopment by saying it is a valuable tool to eliminate blight, generate more tax revenue for public services, and create jobs. The truth? Most redevelopment projects fail to meet their lofty expectations.

There are no guarantees when it comes to economic development. Elected officials—not always known for being good managers of the public's money—must look into their crystal balls and determine whether redevelopment investments will pay off.

Sadly, many politicians have little or no background in finance and simply do not understand the complexities of redevelopment law and its financial intricacies. Therefore, they must rely on the advice of lawyers, consultants, bond brokers, and developers—interests not always aligned with those of the community. Since elected officials often appoint themselves as redevelopment agency members, or as in the *Kelo* case, delegate their powers to private corporations, they gain unparalleled government powers.

Redevelopment agencies can be created without the vote of the citizenry who will be affected. They can incur bond indebtedness without voter approval. They can use the power of eminent domain to take private property and hand it over to favored developers. If that isn't bad enough, they can also line the pockets of those developers with our tax dollars.

Throughout America, there are voices being raised to call attention to this inequity. The Institute for Justice—based in Arlington, Virginia—through its Castle Coalition, calls public attention to eminent domain abuses. Radio talk show hosts Sean Hannity and Neal Boortz, and Fox News Channel's *Hannity & Colmes*, have done an outstanding job in exposing property rights abuses and raising public awareness. Now it is time to expose the framework under which these abuses occur. The first step is to understand the redevelopment process.

FOLLOW THE MONEY

Cities and counties generate most of their income from property taxes and, to a lesser extent, sales taxes. These revenues flow into the municipality's general fund to pay for all government operations and public services, such as the police, schools, and libraries. When elected officials decide there is a need to revitalize a community but there are not sufficient monies in the general fund to do so, they create a redevelopment agency.

The agency designates a specific area it wants to redevelop. In most cases, these redevelopment districts must be declared blighted. Blight is commonly defined as the condition wherein homes or commercial buildings are deteriorating and endangering the public's health, safety, or welfare.

Unfortunately, the definition of blight can be whatever the agency wants it to be. In fact, in Florida there are waterfront homes in gated communities valued at more than $1 million that have been slapped with a blight designation. One only needs to go to the Castle Coalition's Web site (CastleCoalition.org) and click on the "Blighted Home of the Week" page to see how local government has perverted the definition.

Many homeowners fear the blight designation. Property values plummet and it is often extremely difficult to obtain building permits that conform to the agency's plan. Local citizen groups challenge blight designations and are often joined by school districts that stand to lose major property tax revenue if the redevelopment area is approved.

Fighting these designations, however, is an uphill battle. State law and local ordinances can provide powers to the agency's board members that give little or no opportunity for property owners to mount meaningful legal challenges.

Once a redevelopment consultant's blight findings are approved, the agency has the power to expand the redevelopment area. Legal requirements to force a popular vote or referendum to stop this land grab are sometimes difficult to meet. In the vast majority of cases, a vote is never held and the findings of blight are quickly certified.

The agency then receives proposals from developers who paint a rosy picture by declaring that their projects will eliminate blight, create more jobs, and increase property and sales tax revenues that will raise the quality of government services. These redevelopment projects can include "big-box retailers" (such as Wal-Mart, Costco, Lowes, and Target), shopping malls, auto dealerships, sports stadiums, office buildings, hotels, and new housing to replace old neighborhoods.

In order to secure these types of development, enterprising developers often argue convincingly that there is an inordinate amount of market risk and that certain financial subsidies and other incentives are needed for them to justify going ahead with the project. All too often, redevelopment staffs will also tell agency members exactly what they want to hear by using questionable statistics to justify "severe market risk." But if these projects are too risky, as many developers claim, why then should government take the risk?

These subsidies can be in the form of paying contractors to build site improvements that are normally paid by the developer, discounting the cost of the land, or even giving it away for free. This is land that the agency acquires through its powers of eminent domain. In essence, public funds go toward reducing the developer's costs which, of course, increases their profits.

This corporate welfare is not distributed evenly. Favored developers (who have attracted giant discount stores, auto dealers, and others) often receive most of the money while existing small businesses, already burdened by regulations and taxes, must now face larger competitors. What's wrong with this picture?

TAX INCREMENT FINANCING (TIF)

In addition to satisfying their hunger for more property tax revenues, redevelopment agencies also try to attract commercial development that generates more sales tax revenue. Many officials claim that stimulating new sales taxes benefits the city's general fund. On the surface, this is true. In some cases, however, cities find a clever way to pay back the sales tax they generate. It's another giveaway, and here's how it works.

As a way to justify subsidies, many agencies claim that their projects create more jobs. In a few cases this is true. In the majority of cases, however, many new jobs created in redevelopment areas are simply transfers from other areas. As a result, there is no net job gain. In fact, redevelopment projects can result in the loss of jobs outside the redevelopment district since big-box retailers like Wal-Mart can put smaller retailers out of business.

All the job creation figures, financial projections, and incentives to developers can be found in the redevelopment agreement. On close examination of the agreement, citizens can see what tangled webs their elected officials weave. These unholy alliances between agency members, favored developers, contractors, and suppliers would rival most soap opera scripts.

Once the deal is cut with a developer, the agency must finance the proposed project. By law, since money from the city's general fund cannot be used for redevelopment, these projects are financed by the city selling bonds to the public through a mechanism known as tax increment financing (TIF).

For example, a developer convinces a city that bulldozing a neighborhood of older homes and small businesses and replacing it with a retail and office complex will generate $500,000 in annual property tax revenues. Assuming the "old" neighborhood already produces $100,000 of property tax revenue, the city can incur bond indebtedness in an amount equal to the difference, or $400,000.

By law, the city must still receive $100,000 of tax revenues even after the neighborhood is gone. In theory, the "projected" future tax revenues generated by the new development go toward paying the bond debt. After the bonds are paid off, the city will then start receiving all $400,000 of tax revenue.

Although TIF, with the helping hand of eminent domain, can help revitalize our inner cities, the evidence is *overwhelming* that many redevelopment projects fail to meet their projections. There are, however, redevelopment projects that have been successful without seizing property through eminent domain.

SMART ALTERNATIVES

In Bonita Springs, Florida, the city, in conjunction with the National Trust for Historic Preservation and the Department of Housing and Urban Development, purchases abandoned and foreclosed homes that have been on the market for more than six months. The properties are then renovated and sold to low-income or moderate-income families. These homeowners can then restore their exteriors once a property is designated a "Beautify Bonita" project. Tax-deductible donations are solicited and volunteers make the repairs—all this without the use of eminent domain.

In the late nineties, three blocks in the heart of Seattle were in need of revitalizing. Properties were acquired and redeveloped into an upscale retail complex without the need to condemn. Investors used land swaps and street maintenance agreements through the city to attract major stores to key corners. This paved the way for boutique shops and establishments to fill the remaining space.

There are also examples where the limited use of eminent domain contributed to the success of the redevelopment plan. One such case was outside Indianapolis. A neighborhood just north of downtown, now known as Fall Creek Place, was severely blighted and known for its violence and drugs. One thing that complicated changing these conditions was the fact that several homes had been abandoned and vacant lots had been poorly maintained.

The city embarked on a plan to acquire 250 properties of which only twenty-eight were eminent domain cases. According to Indianapolis Mayor Bart Peterson, the city did not use eminent domain against any property owner's will—except when property owners could not be located.

Today, Fall Creek Place is a mixed-income neighborhood with homeowners from all backgrounds. The majority of the residents are in the low-income bracket, and more than 70 percent are first-time homeowners. The project has spurred private development in the area. In fact, construction of "live-work units" that feature retail stores on the first floor and residential space above is in the works.

Unfortunately, the Bonita, Indianapolis, and Seattle cases are exceptions to the rule. The redevelopment industry continues to thrive and, in many cases, perpetuates a culture of short-term expediency and greed rather than protecting the long-term economic health of communities. Why? Because the elected officials, lawyers, bond brokers, consultants, and developers are long gone when the bond debt comes due twenty or more years later.

Public officials often point the finger at previous administrations when they are forced to refinance debt, which is like transferring the balance from one credit card to another. Another ploy is to siphon off money from other redevelopment districts. Some cities may even tap into their general fund. This reduces money for public services, and inevitably the cities are forced to increase taxes.

In the long run, who ends up paying for these redevelopment scams? You, the taxpayer.

4

A LITTLE PINK COTTAGE BECOMES A NATIONAL SYMBOL
(New London, Connecticut)

"[The Kelo case] has to be one of the most vilified Supreme Court decisions in history."
—SCOTT SAWYER, A NEW LONDON, CONNECTICUT, LAWYER

Seven property owners in a small Connecticut town could never have imagined that their court battle would ignite a national debate. *Kelo v. New London*, a landmark U.S. Supreme Court case, shocked millions of Americans who now realized they, too, could have their homes and businesses seized by the government and handed over to private developers.

New London, a city on Long Island Sound, in the southwestern part of the state, had fallen on hard times in the late 1990s. The town had experienced serious declines in employment, especially after the loss of 1,900 jobs when the U.S. Naval Undersea Warfare Center was closed in 1996. The small, blue-collar community was dying a slow death. All that changed in February 1998.

Connecticut Governor John Rowland and New London Mayor Lloyd Beachy then announced that Pfizer, one of the largest pharmaceutical manufacturers in the world and the makers of Viagra, would build a new 400,000-square-foot research and development facility in the city. The new project would employ 2,000 workers and was expected to generate several hundred jobs in the community as well as badly needed tax revenues. It all looked great on paper.

Pfizer proposed to build their campus on twenty-four acres on the old New London Mills site in the Fort Trumbull area of town. Here was the deal:

- The property would be purchased for a mere $10;
- Pfizer would be exempt from property taxation for ten years; and
- Pfizer would receive some $118 million in state and federal assistance.

In addition, since the soils on the property had to be cleaned from the toxic waste generated by industries that had closed in the late 1970s, the city was ridding itself of a messy problem as well as getting an economic boost.

LET THE GAMES BEGIN

To capitalize on the project, New London received approval from the state to sell bonds to redevelop ninety acres adjacent to the Fort Trumbull State Park, including the former naval site. The goal was to create more development that would complement the Pfizer facility and provide public access to the city's waterfront, "building momentum" for the rest of the community.

The proposed development would include new housing, office buildings, and retail space in addition to a luxury hotel and conference center, a health club, and a U.S. Coast Guard Museum. Overseeing the plan would be the New London Development Corporation (NLDC), a private, nonprofit development company.

It is commonplace for cities to form redevelopment agencies to oversee revitalization projects. Traditionally, elected officials appoint themselves to serve on the boards. However, the NLDC, headed by Claire Gaudiani, President of Connecticut College and wife of a Pfizer executive, was composed of local business and civic leaders who were authorized to use the city's eminent domain powers. Clearly, this unusual arrangement presented a danger for potential conflicts of interest and the compromising of the public interest.

It is clear from later court testimony that the NLDC and Pfizer were at work long before the deal was announced. A December 1997 letter from Claire Gaudiani to the head of Pfizer's research division stated that the NLDC would work with the company to relocate a scrap dealer, upgrade utilities services in the area, and acquire a number of surrounding properties. Evidence also showed that some property owners in the designated area began to be intimidated by realtors who aligned themselves with the NLDC.

It was also clear that the City of New London planned to take a backseat role in the process. Mayor Beachy would later testify that the whole project was "a sham" and that the Connecticut Department of Economic and Community Development (DECD) dictated every action the city would take.

While the Pfizer-approved plan was being developed, the city council held executive sessions—meetings not open to the public—and were given resolutions by DECD representatives to sign and pass. No changes were permitted.

On at least one occasion, Mayor Beachy met with James Abromaitis, a DECD commissioner, and other state officials in Hartford to protest that the city was having the Pfizer project "shoved down its throat." During the two-hour meeting, the mayor asked what role the city would play in the redevelopment process. Beachy claims Abromaitis responded, "Mr. Beachy, you have to understand, it's our money and we're going to do what we want with it."

NEIGHBORS UNDER SIEGE

In early 2000, plans were unveiled on the Fort Trumbull re-vitalization plans. Among the property owners within the redevelopment area was Susette Kelo.

Three years earlier, while working as a paramedic, Su-sette searched in the Fort Trumbull area and found a little Victorian cottage built in 1895 with beautiful views of Long Island Sound. The home was in shambles and was so over-grown with weeds that Susette needed a machete to reach the front door. She scraped together enough money to buy it and spent her spare time fixing it up to create the kind of home she had always dreamed of having. She even painted it salmon pink, her favorite color.

Within a year after Susette had bought the property, a real estate agent stopped by and made her an offer on be-half of an unnamed buyer. Susette said she was not inter-ested in selling. The realtor said that her home would be taken by eminent domain if she refused to sell and told her to give up because the "government always wins."

For months, Susette worried about what would hap-pen next. On the day before Thanksgiving 2000, the sheriff taped a letter to her door. Her home had been condemned by the NLDC and she was ordered to vacate by March or police would forcibly remove her and her belongings.

One month later, Scott Sawyer, a local attorney, in con-junction with the Institute for Justice, a nonprofit legal foundation, agreed to represent Susette and her neighbors and filed suit in superior court to stop the evictions.

Among Susette's 'neighbors joining in the suit was Matt Dery. The close-knit Dery family had been residents of New London for more than a hundred years. Matt, his wife Su-zanne, and their son lived next door to Matt's mother and father in a house that Matt's great-grandmother had pur-chased in 1901.

The Derys were not only residents but part of the community's rich and historic past. Matt's grandmother had opened a grocery store on property that was now being seized and had extended credit to everyone in the neighborhood when they needed it. When her own home went into receivership during the Depression, she worked until 1958 to earn it back. Now, the NLDC had decided it wanted someone else to live there.

Around the corner from the Derys was Bill Von Winkle's Fort Trumbull Deli that had served oversized hoagies since 1986. Bill and his wife Jennifer owned six apartments above the deli, which they leased to tenants. In the area, the Von Winkles also owned two homes, five additional apartments, and a commercial building with three storefronts, all of which were leased to tenants.

The NLDC wanted it all and even padlocked one of the apartments—while a tenant was still inside! That same day, in freezing mid-January, the NLDC also forced tenants out into the street in their stocking feet.

Not far from Von Winkle's deli, demolition work had already begun on older homes in the neighborhood. Next to a dusty construction site lived Jim Guretsky, his wife Laura, and their two young daughters. The Guretskys owned a three-unit apartment building. They lived in one unit and rented out the other two units to tenants.

One day, some heavy equipment slammed into their kitchen. Luckily, Laura and a daughter escaped injury. Fearing for their health from the constant dust and hazardous material (such as lead and asbestos) swirling through the air, the family decided to move to Pennsylvania until things settled down.

A block from the Guretskys were two houses that Richard Beyer and a partner had bought in 1994. They had seen promise in the neighborhood and had worked thousands of hours to renovate the properties. Dumpsters full of trash were removed. In a parking lot area between the homes, 15,000 bricks had been laid by hand. Granite and other stone was also hand cut to spruce up the interiors.

According to court testimony, after Richard was given notice that their properties were to be seized, contractors vandalized the home by tearing up hand-bent metal fittings around each of the windows and using claw hammers to dig into the new siding. When he and his partner contacted the police to file a report, they were threatened with arrest and told that they we were trespassing on NLDC property.

Next door to one of the homes lived Pasquale and Margherita Cristofaro. As with Susette Kelo, a sheriff's deputy showed up at their front door on a day before Thanksgiving with condemnation papers and ordered them to vacate by March. Margherita began having heart palpitations and had to be taken to the hospital.

This was the second time a member of the Cristofaro family had had their home seized by the city and handed over to a private developer. In the early 1970s, New London had seized a home to build a sea wall. Instead of building the wall, a private development was built instead.

SUCH A DEAL!

Had it not been for the intervention of the Institute for Justice, Susette Kelo and the other property owners would have been unable to afford the high legal fees required to sue the city. Although they all knew it was an uphill battle, they would have their day in court.

During their seven-day trial in early 2002, a stunning announcement was issued by the NLDC that was seen to pressure the "holdouts" to settle. The NLDC announced that it had chosen Boston-based developer Corcoran Jennison to build a 110-room hotel and conference center and apartments on the abandoned U.S. Navy property. In addition, a planned biotechnology office complex would be built in the neighborhood.

Pfizer, it was reported, would commit to use the hotel and conference facilities and the state would chip in $75 million to build the streets and utility services—as well as to purchase and demolish homes. The deal, according to Admiral David Goebel, the chief operating officer of NLDC, was that the land would be owned by the NLDC and then leased to Corcoran for ninety-nine years. The price tag? One dollar!

The NLDC's public position was that the development plan would have a "significant socio-economic impact" on the city and was expected to generate around 800 construction jobs, up to 1,300 direct jobs, and approximately 900 indirect jobs. What wasn't highly publicized, however, was that more than half the property would be exempt from taxation.

The remaining property, it was estimated, would generate between $680,000 and $1,200,000 in annual property tax revenues for the city. Best case scenario: The $75 million investment would be recouped in sixty-two years.

These projections were based upon a certain amount of square footage being built. However, there were no assurances of how much or when this development would take place; nor was there any penalty if the developer did not perform as expected.

One property in the area that escaped the clutches of the NLDC, even though it was next to one of the private homes being seized, was the Italian Dramatic Club. Reportedly, the club was a watering hole for local politicians and was frequented by Governor Rowland (later convicted in a corruption scandal and sent to prison).

*   *   *

A BLANK CHECK

Finally, on March 13, 2002, the Superior Court of New London rendered its decision. The judge dismissed eleven of fifteen eminent domain actions on certain parcels within the redevelopment area—but not the area where Susette Kelo and the others lived. The court ruled that economic development constituted a valid public use and that the takings would "sufficiently benefit the public and bear reasonable assurances of future public use." Both sides appealed the case to the Connecticut Supreme Court.

On March 23, 2004, came the Connecticut Supreme Court's stunning decision. In a 4-to-3 vote, the court ruled against the property owners. Among the arguments that the three dissenting justices raised was that the city did not prove its redevelopment plan provided enough public benefit to warrant the extreme measure of taking private homes. Justice Peter T. Zarella wrote, "The tremendous social costs of takings . . . are difficult to quantify but nonetheless real. The fact that certain families lived in their homes for decades and wish to remain should not be . . . dismissed."

The decision was, of course, very disappointing, but the Institute for Justice and the property owners vowed to continue the fight. Scott Bullock, a senior attorney at the institute, was quoted as saying, "If allowed to stand, this decision gives local officials a virtual blank check to condemn private property at the whim of private parties."

A motion was filed for the court to reconsider its ruling. Bullock wrote: "Sacrificing fundamental constitutional protections because of a perceived pressing need is both bad policy and bad law. As long as government jumps through all the procedural hoops, it can freely condemn property for private businesses that, as long as they make an ordinary profit, will probably generate more taxes than those pesky low-tax homes."

On April 20, 2004, the Connecticut Supreme Court declined to reconsider the case but issued a "stay" to stop the city from evicting the owners until the U.S. Supreme Court either declined to hear the case or agreed to accept it.

TO THE HIGHEST COURT IN THE LAND

On July 19, 2004, the Institute for Justice filed its appeal in the U.S. Supreme Court. In addition, twenty-five *amicus curiae*, or "friends of the court," briefs were filed in support of the property owners. Among them were the National Association for the Advancement of Colored People (NAACP), the American Association of Retired People (AARP), the Cato Institute (a conservative think tank), the American Farm Bureau and the Farm Bureau Federations of eighteen states, the National Association of Home Builders, and the National Association of Realtors.

Clearly, the *Kelo* case had sparked the nation's attention and that of the nine justices of the Supreme Court. They agreed to hear the case. But on June 23, 2005, the high court, in a narrow 5-to-4 vote, ruled against the property owners. Susette Kelo and her neighbors were handed their third and most crushing defeat.

One of the key questions in the case was whether economic development is distinguishable from more traditional uses of eminent domain—to build roads, schools, dams, and other "public use" projects. The majority opinion, written by Justice John Paul Stevens, stated that there was no difference based on case law precedent. In fact, he wrote that "the court declines to second-guess the wisdom of the . . . city" in such matters.

In a dissenting opinion, Justice Sandra Day O'Connor wrote: "Today the Court abandons this long-held, basic limitation on government power. Under the banner of economic development, all private property is now vulnerable to being taken and transferred to another private party, so long as it is upgraded." She added, "Nothing is to prevent the State from replacing any Motel 6 with a Ritz-Carlton, any home with a shopping mall, or any farm with a factory."

Except for Mike Cristofaro and Susette Kelo, the five other property owners agreed to settle and vacate their homes before a May 31, 2006, deadline.

Within a few weeks after the deadline, the Cristofaros finally agreed to leave their home after the city agreed to support more housing in the Fort Trumbull area and give the family exclusive right to buy one at a fixed price. In addition, the city agreed to install a plaque in the neighborhood commemorating the loss of matriarch Margherita Cristofaro, who had passed away during the court battle.

Faced with eviction and the destruction of her home, Susette Kelo made an offer to the city, as she had done years earlier, to preserve the home and move to another location close to Long Island Sound. The city agreed.

The fight of those seven property owners that had begun eight years earlier had come to an end. As Susette Kelo so aptly said before a Senate hearing: "This battle against eminent domain may have started as a way to save my little pink cottage, but it has rightfully grown into something much larger—the fight to restore the American Dream and the sacredness and security of each of our homes."

THOSE PESKY HOLDOUTS
(Norwood, Ohio)

"In this country, government isn't supposed to be a super real estate agent, throwing out rightful land owners to make way for the rich."

—DANA BERLINER, SENIOR ATTORNEY, INSTITUTE FOR JUSTICE

The decision of the City of Norwood, Ohio, to "rent out" its powers of eminent domain to a developer landed it in court. The battle had national implications and serves as a stark reminder of how far cities—hungry for tax revenue—and developers will go in the name of *urban renewal*.

Norwood, a small town with a population of about 21,000, is encircled by the city of Cincinnati. For years the city's economy had been healthy until a General Motors assembly plant and a tool manufacturing company closed down. As a result, the city felt it needed to revitalize certain areas, create more jobs, and increase tax revenues to provide better government services. One of the best ways to accomplish this is through commercial development. Norwood had an ideal location just five minutes from downtown Cincinnati.

In the late 1970s, Interstate 71 was built through the southeast edge of Norwood. The town created a redevelop-

ment district, known as the Edwards Road Corridor Urban
Renewal Area, to attract development.

PROMISES OF RICHES

One of the first major developments in the district was the
Rookwood Commons & Pavilion. The mixed-use project,
built by Jeffrey R. Anderson Real Estate (a successful commercial developer with real estate assets estimated to be
$500 million), included office space and a twenty-three-
acre shopping center with forty-nine stores such as The
Gap and Ann Taylor.

Eager to capitalize on the success of the project, Anderson then sought to expand its holdings outside the
renewal area. Plans were announced to build Rookwood
Exchange, a retail center anchored by Crate & Barrel, a
home goods chain. The property was also designed to offer office space—more than double that of the Rockwood
Commons—as well as 200 upscale condos.

Based upon the developer's forecasts, the $125 million
project was estimated to increase the city's tax revenues
between $1.5 to $3 million, which represented 15 to 20
percent of the city's annual budget. Council members were
ecstatic. Standing in the way of the sprawling project, however, was a middle-class neighborhood of ninety-nine modest homes and businesses.

In 2002 Anderson began its campaign to sell the project to the city. It made presentations to the redevelopment
committee, its planning commission, and the city council. Several public hearings on the project were also held.
According to court records, from the outset Anderson re-
peatedly pressed the city to invoke its eminent domain
powers.

Each request was denied and the developer was urged

to acquire the properties through private negotiation. In fact, the city required Anderson to document its acquisition efforts, which began in the summer of 2002.

At first, about half of the property owners rejected Anderson's purchase offers. One resident, a 35-year-old construction manager named Joe Horney, who had bought a two-family house with an inheritance when he was 21, circulated a petition around the neighborhood. He found that most of the residents wanted to stay. Included in that group were Carl and Joy Gamble.

The Gambles had worked hard all their lives. Their home of more than thirty-four years was well kept and had a large backyard with a garden. They had raised their two children in the modest house, and after selling their family-owned grocery store, had planned to live out the rest of their lives there quietly gardening.

Mary Beth Wilker and her husband Nick Motz were also against the project. They had bought and renovated a building in the neighborhood where Beth operated her graphic design business, Wilker Design.

JUSTIFYING BLIGHT

By September 2003, however, Anderson and his partners announced they would buy the properties for 35 percent higher than fair market value. One catch: *everyone* had to sell or the deal was off. This is a common tactic used by redevelopers and inevitably causes friction among property owners.

About a dozen owners, including Horney, the Gambles, and Wilker stood firm. The property owners who agreed to sell either thought the offer was generous or realized the city would eventually win anyway. With the majority of the neighborhood willing to sell, the next step was to undertake an urban renewal study that would—among other things—justify the use of eminent domain, if necessary.

In December, Anderson approached the Norwood City Council to undertake the study and admitted its "findings are key only if we can't get the property owners to sell." In other words, the report would have to show that the neighborhood was blighted.

The definition of the term "blight" varies widely. Cities can (and have) created definitions to suit their own purposes. The traditional meaning of a blighted area refers to homes and businesses that are run-down or deteriorating to the point that they endanger the public's heath, safety, and welfare. In other words, a slum. As in the Norwood case, the definition has been frequently broadened to assist a private developer to acquire properties from unwilling sellers with the threat or use of eminent domain.

The city agreed to undertake a study. Anderson agreed to pay for it and an architectural and urban design consultant was hired. In the meantime, property owners who wanted to sell were in limbo.

THE BOGUS STUDY

The survey by the consultant found that the construction of Interstate 71 had had a negative effect on the neighborhood. It suggested that the highest and best use of the area was for commercial development rather than residential. It also found that development pressure, consumer demand, and the fact that the majority of the owners wanted to sell indicated that "piecemeal redevelopment would occur." This, they concluded, would have "an adverse effect on the physical, aesthetic, and functional qualities of the area."

It was clear the study was intended to show that the neighborhood was blighted, and it was fueled by the desire to move the private development forward. In order for the area to "qualify," the consultants relied on factors usually not associated with what would normally be considered slum conditions.

The study cited such things as broken pavement along sidewalks, standing water on roads, and poorly designed streets—none of which the owners could control. Although these purported blight conditions existed, it was the city that had either created them or had allowed them to persist.

The study found no dilapidated homes or businesses. However, several conditions cited in the study exaggerated the impression that the condition of the neighborhood was deteriorating. The report noted weeds in front yards, cars parked on front lawns, and a lack of required safety rails and hand railings. Several of these items were double- and triple-counted, rendering the study's credibility questionable. The property owners against the sale attacked the study vigorously.

The Institute for Justice decided to represent the owners. It had been working on an eminent domain case in Lakewood, Ohio, in which the city was using similar tactics to justify a blight designation to make way for private development.

One blight factor used in both the Norwood and Lakewood studies was the lack of attached two-car garages or homes with fewer than two full bathrooms. Coincidentally, Anderson was one of three companies vying for the Lakewood project.

THE CITY GIVES THE GREEN LIGHT

During the August 2003 city council meeting, Scott Bullock of the Institute for Justice said, "If the city authorizes this study, it would, frankly, be a fraud." But despite several objections to the study, the city council unanimously passed an emergency ordinance "to eliminate deteriorating and deteriorated areas" within the city "and improve safety and traffic conditions and other deteriorating conditions in the area" to encourage "prompt redevelopment."

City Councilwoman Cassandra Brown, who represented the ward where the neighborhood was located, said: "We have an obligation to do what is best for all the people of Norwood," and, joining other council members, stated that approving the study didn't necessarily mean that the city would use eminent domain to buy out the remaining owners.

The city council then passed an ordinance authorizing the mayor to contract with Rookwood Partners, a partnership formed by Anderson, to undertake the project. The contract specified that Rookwood was to pay all costs associated with the eminent domain process and also required it to demolish all the existing buildings in the renewal area and improve the road alignments for circulation.

In September, the council passed a resolution declaring the intent to exercise its eminent domain power over the property owners. A month later, it reluctantly filed condemnation actions in the Hamilton County Common Pleas Court. Mayor Tom Williams said, "I was a cop for thirty-four years, got shot once, and shot people twice. It's the same thing with this thing. You hate to pull the trigger, but sometimes it's a necessity."

The hopes of property owners wanting to sell were bolstered. The Institute for Justice, Joe Horney, the Gambles, and others vowed to fight the condemnation order. They filed a lawsuit challenging the blight designation, citing several deficiencies in the study and alleging that the city was not following its own policies.

On the first day of trial—an unseasonably cold winter's

day—a rally was held by a large group of people who had assembled on the courthouse steps. American flags were being waved and several anti–eminent domain banners were held high. Inside the courtroom, the crowd was so large that the bailiff had to demand that dozens of people wait out in the hall.

EVICT AND DEMOLISH

The emotional high point of the trial came when Joy Gamble took the stand. Bullock, who would later say that Joy reminded him of his grandmother who had passed away a few years earlier, found it difficult to ask her questions.

At the end of the testimony, Bullock asked why she wanted to keep her home. She replied, "All of our memories are there. We're rooted there. We don't want to be uprooted. We want to stay right here until we're carried off feet first." The audience was clearly moved.

The city's attorney, who had engaged in aggressive questioning of other property owners earlier, rose and walked to the podium to cross-examine Mrs. Gamble. He paused, looked at Mrs. Gamble and then the judge, said, "No questions," and sat down. Bullock later wrote that "he must have felt a collective vibe in the courtroom that implied: *'Don't you dare go after her!'*"

The trial court judge, Beth A. Myers, agreed with Horney, the Gambles, and other owners that the city council had "abused its discretion in determining the renewal area was a slum, blighted, or deteriorated." However, she sided with the city and said the area was "deteriorating" and cited a provision in the city's code that "diversity of ownership" also contributed to the blighted condition of the area. In other words, multiple owners. What neighborhood in America does *not* have multiple owners?

Myers dismissed the suit of the Institute for Justice and ruled that the city could use its eminent domain powers to transfer the five remaining homes to the developer. Dana Berliner, a senior attorney for the Institute, was disappointed with the court's ruling but said, "The City of Norwood is on much shakier legal ground in justifying the use of eminent domain because a neighborhood is supposedly deteriorating. The Ohio Supreme Court has never held that a city can take land under the incredibly broad criteria of 'deteriorating' and we look forward to presenting the vital issue to the appellate courts."

Among the issues raised in the appeal was the constitutionality of the city's use of eminent domain. Lawyers for the institute argued that the Fifth Amendment of the U.S. Constitution, which states that private property can be taken only for "public use," prohibited the city from using its powers of eminent domain to condemn property for private development. The institute also cited provisions in Ohio's constitution to support its position.

On September 3, 2004, the appeals court ruled in favor of the "holdouts" and said that Horney and the others could bring a constitutional claim against the city. The court also reinstated the original suit that Judge Myers had dismissed a year earlier. The legal battle continued.

In other court actions a few months later, Judge Myers ruled that the developer was free to evict Horney's tenants and the Gambles *and* demolish their homes while the constitutional issue was still making its way through the courts. Although the judge said that the developer assumed the risk that Horney and the Gambles could get their properties back on appeal, it was of little consolation.

Scott Bullock said that Judge Myers' ruling was "indefensible" and that "private developers can not only take your home and business with the city's help, they can destroy it before your appeal is heard. That's outrageous!" An appeal was filed immediately.

In a shocking reversal of its earlier ruling, the court of appeals sided with Judge Myers this time and said the city could evict and demolish the properties before a final resolution of the property owners' constitutional claims was reached. The institute then filed a motion with the Ohio Supreme Court to stop the evictions.

A COMMUNITY AT ODDS

Although Horney and the Gambles vowed to continue the fight, the Gambles could no longer withstand the constant strain and uncertainty. They opted to leave their home in the middle of winter and moved into the basement of their daughter's home in Kentucky.

In February 2005 the Ohio Supreme Court ruled that the homes must be protected while the Ohio courts considered the appeals. The case was then sent back to the appeals court to determine if the owners would get their homes back.

By this time, more than two years had elapsed since the property owners had first filed their suit. The legal costs of the Institute for Justice were approaching $500,000. The other property owners—the majority who had already made their deals with Anderson—were becoming increasingly irritated with the holdouts. Things had begun to get ugly.

One property owner who wanted to sell out and get on with his life told a journalist that several people were mad and disgusted "because they ain't got their money yet." Another frustrated owner told a reporter: "F*** the Gambles! Yeah. F*** the Gambles . . . and throw Joe Horney in there, too!"

Yellow signs sprouted up throughout the neighborhood that read "Held Hostage by the Institute for Justice" and

"We Support Rookwood Exchange," while property owners supporting Horney and the Gambles placed signs on
their properties with the word Eminent Domain inside a
circle with a red line through it.

The legal battle was not only creating financial and
emotional hardships; it was also visibly taking its toll on
the neighborhood. With demolition imminent, some owners moved out, rented their homes, or left them empty and
simply stopped maintaining them.

The case returned to the appeals court. Again, the court
ruled that the city could seize the remaining properties.
However, the injunction imposed by the Ohio Supreme
Court to stop the homes from being destroyed until all appeals had been exhausted, remained in place.

The Institute for Justice team petitioned the Ohio Supreme Court to hear the case again. They argued that the
city had not followed its own ordinances and that the taking was unconstitutional under the Fifth Amendment of
the U.S. Constitution and also under the Ohio constitution.
Among the other arguments they made was that the urban
renewal study was flawed. In fact, 47 percent of the factors
used to determine that the area was deteriorating were in-
correct.

While the state high court considered hearing the case,
Kelo v. New London, Connecticut was heard before the U.S.
Supreme Court. The Institute for Justice also represented
the property owners in that case and the issues between it
and the Norwood lawsuits were similar: a city using its powers of eminent domain to take nonblighted private property
for the purpose of increasing tax revenues. Both sides in
the Norwood case knew that the *Kelo* decision could have a
direct influence on the outcome of the case.

Just when it seemed that the friction in the community
could not get worse, the city spray-painted a large "X" on
the front door of the Gambles' home. A tree company also

marked trees for loggers to cut down. The Gambles, who were not even allowed to enter the premises, were devastated.

The Ohio Supreme Court agreed to hear the Norwood case in May 2005, but on June 23, 2005, the U.S. Supreme Court issued the *Kelo* decision. The ruling: 5 to 4 in favor of the City of New London, Connecticut.

Clearly, the ruling was a blow to Horney and the Gambles. However, the U.S. Supreme Court ruling left the door open so that states could place "further restrictions on the exercise of the takings power."

HOME SWEET HOME

On July 26, 2006, the Ohio Supreme Court ruled unanimously that the City of Norwood could not use eminent domain to take the homes of Carl and Joy Gamble and Joe Horney.

The Ohio court rejected the U.S. Supreme Court's decision in *Kelo* that local governments can take property from one party and transfer it to another simply to produce more taxes. It also said that state courts must use "heightened scrutiny" when property is taken by another private party.

"Our home is ours again!" exclaimed Joy Gamble. "The Ohio Supreme Court has stopped this piracy." Joe Horney said, "I am so excited! Wow! I can't wait to see my old place. I feel like giving it a big hug!"

One day after the ruling, the Gambles and Joe Horney went to what had been their old neighborhood. There, in the middle of a grassy field stood their homes.

They hugged one another in front of a chain-link fence and then, with members of the media, walked toward their homes.

Joy noticed that their $1,000 fence had been stolen.

It was also apparent that the home was not in move-in condition, since the utilities had been cut off. When she opened her front door marked with an "X," she saw that the house had been looted. "This house was entrusted to the developer," she said. "He should have seen that nothing was done. He didn't care. He just thought that the Ohio Supreme Court would rule in his favor."

Despite the damage, she and her husband were relieved to have their home back. Justice, at least in this instance, had prevailed.

6

MURKY WATERS IN RIVIERA BEACH
(Riviera Beach, Florida)

"The people who live on the water are cheating the poorest members of our community."
—RIVIERA BEACH MAYOR MICHAEL BROWN

The term "negro removal" came about in the 1950s and 1960s because urban renewal projects targeted predominantly poor, minority neighborhoods. There were charges of racism, clashes between the haves and have-nots, political bickering, and accusations that cities had sold out to big money developers. Fast-forward to the present, and much of the same is being said about a small Florida town.

Although the city of Riviera Beach sits across an inlet from its Atlantic coastline neighbor, West Palm Beach, there are many stark differences. West Palm, considered to be one of the wealthiest cities in the state, is predominantly white. Riviera Beach, on the other hand, is one of the poorest in Florida and has an African-American population of nearly 70 percent.

One-quarter of Riviera Beach's residents live in poverty. The 2000 Census showed that one out of every four homes had three rooms or less—a figure normally associated with overcrowding. In addition, 80 homes in the community had no plumbing and 327 had no source of heat.

AN AMBITIOUS PLAN

Despite the high unemployment and crime rates, the city's
waterfront location made it ripe for development. Since
much of the south Florida coastline had been lost to high-
end residential projects, the demand for marinas and dock
space was outstripping supply.

The city embarked on a redevelopment effort in the late
1980s. Progress was slow, but by 1999 a 400-acre Community Redevelopment Area (CRA) was formed, which reportedly included 2,262 households and 317 businesses.

The city's redevelopment agency, comprised of the city
council members, had also reported that there were more
than 1,000 renters and only 342 actual homeowners in the
CRA. In fact, a councilwoman was quoted as saying that
the reason the city was blighted was because of the inatten-
tion of absentee landlords—a politically correct term for
"slumlords."

By some accounts, the plan would be the largest revitalization project in the country and could displace between
5,100 and 6,000 people through the use of eminent domain.
Mayor Michael Brown, an African-American attorney and
Riviera Beach native who is credited with spearheading the
redevelopment effort, said it was necessary to use "tools
that have been available to governments for years to bring
communities like ours out of the economic doldrums and
the trauma centers." He added, if we "don't use this power,
cities will die."

Plans to resuscitate the city began with an urban renewal plan. Although many property owners saw the plan
as a way to sell their homes and businesses, not all residents were in favor of it. One such homeowner was Martha
Babson.

Babson, a self-proclaimed "old hippie" and community
activist, had lived in Riviera Beach for more than twenty-

three years. The green cottage she shared with her dog and parrots was perched on the Intercoastal Waterway—a prime location.

Babson challenged the 2001 urban renewal study. After she discovered several errors in the report, two property owners—an owner of a boat service company and a Singer Island café owner—scraped together $350 and paid her to document the mistakes.

Equipped with a borrowed camera, clipboard, pen, and the CRA's color-coded map, Babson walked down every street and did a parcel by parcel investigation. She found that there were several homes listed as vacant lots and that houses in good condition were designated as being dilapidated. Her overall conclusion: "It looked like [the study] had been done by two guys sitting in a bar and saying, 'Let's throw this in.'"

Babson offered her findings at a June 2001 city council meeting. The pro-redevelopment council didn't take her report seriously. She wrapped her paperwork in plastic and put it on her kitchen shelf.

THE DOG-AND-PONY SHOW

Over the next few years, the plans for the revitalization project seemed to sputter. When the city began its effort to attract developers, however, residents in the redevelopment area started asking questions.

One couple, David and Rene Corie, who had been residents for five years before the study was funded, kept asking the city what it planned to do with their home. When they were finally told that the developer of the project would decide what to do with their property, Rene said, "That really set me off!"

With its waterfront location and the political will to revitalize the community, Riviera Beach became a prime development target. One city councilman, Edward Rodgers, compared the city to a pretty girl with problems that all the fellas think they can deal with if they get a date. The developers "feel they can polish us up and make us something nice," he said, "and I believe them."

Five suitors came calling. Reportedly, each developer spent around $1 million on lavish presentations. Some residents on Singer Island, however, were not impressed.

Singer, an upscale and mostly white community that had a history of being at odds with the mostly black city council, told the council not to be dazzled by all the pretty pictures and graphs and pie-in-the-sky projections. Councilman Rodgers, however, was smitten and said that it was clear that the council was not "any competition for any of those developers."

On September 14, 2005 (three months after the *Kelo* ruling), the city chose to start negotiations with Viking Inlet Properties to serve as the CRA's "master developer." Viking Inlet was a partnership between Viking Group, a New Jersey-based boat building company, and Portfolio Group, a resort developer based on Singer Island that had developments throughout Australia.

The selection made sense. Viking operated a boat service yard in the area that employed ninety people. The company wanted to relocate three of its divisions to Riviera Beach—its high-end yacht and sports cruiser company, a marine electrical service, and a manufacturing company that made fish spotting towers.

The chairman of Viking, Robert Healy, estimated that the moves would add 750 jobs in three to five years. He also made the point that Viking was not just another builder who was just going to develop a piece of property and then leave. "We're part of the community," Healy said, "and we're going to stay part of the community."

Another factor that had weighed heavily in the council's selection of Viking Inlet was the company's pledge to help aerospace giant Lockheed Martin expand one of its nearby divisions that made unmanned undersea vehicles and remote mine-hunting systems. The expansion reportedly would add 150 additional jobs.

A MEDIA FRENZY

Since it was projected that more than 1,000 jobs would be created by the revitalization project, Healy recognized that the city needed a trained workforce. Therefore, Viking sponsored a charter school for more than 300 high school students who were interested in working in the maritime industry. The estimated cost to the company to open the school in the first year alone would be about $500,000, but Healy stressed, "If you want to build good boats, you've got to build good people."

The scope of the revitalization project was immense. It was estimated to cost $2.5 billion and span ten to fifteen years. One of its more ambitious plans was to move U.S. Highway 1 and dig a man-made lagoon to dock large yachts. The project also involved acquiring several existing waterfront properties—through eminent domain if necessary—to build condos, office space, a 250-room hotel, a multi-story boat storage, and even a large aquarium.

Due to the *Kelo* ruling and other eminent domain abuse cases, the project received a lot of media attention. Among the commentators who weighed in were Sean Hannity, the conservative radio talk show host, and his liberal counterpart Alan Colmes on Fox News Channel's *Hannity & Colmes*. The two men rarely agree on issues but had become harsh critics of the government's use of eminent domain to generate more tax revenue.

During a December 2005 special edition of *Hannity & Colmes*, broadcast from Riviera Beach, Hannity interviewed a number of homeowners and business owners who opposed the redevelopment scheme. One of the more compelling clips from the show, which took place in front of a home under the threat of condemnation, was a heated exchange between Hannity and Mayor Brown.

HANNITY: Are you going to kick them out? Answer this directly. Yes or no? Are you going to kick them out?
BROWN: We are going to rescue and relocate.
HANNITY: Are you going to kick them out?
BROWN: And we will put them in a better position.
HANNITY: And if they don't leave, what are you going to do?
BROWN: Well, then the—look, it's the basic tenet of American jurisprudence . . .
HANNITY: To kick them out?
BROWN: For the public service and for the good of the community, sometimes . . .
HANNITY: To give to another citizen.
BROWN: . . . you have to make—no, not to another citizen.
HANNITY: A developer.
BROWN: It's a redevelopment plan. It comes to the city. It comes to the government.
HANNITY: And then a developer.
BROWN: And then that plan—it's no different than a building or a courthouse or a hospital or anything else.

Most Americans, and the Institute for Justice, disagreed.

* * *

WAGING BATTLE

Fresh from its stunning victory before the Ohio Supreme Court in the Norwood case, the Institute for Justice saw an opportunity. Through a media blitz, it reported that Florida had one of the weakest protections against eminent domain abuse. It then joined with the prominent Florida law firm of Brigham, Moore, the Pacific Legal Foundation, and other property rights groups, to aggressively lobby Governor Jeb Bush and the legislature to reform the law.

With all this media attention, Riviera Beach property owners like Martha Babson were reenergized. Babson dusted off the dog-eared survey she had done in 2001, and lawyers began building a case.

In addition to errors in the number of homes and vacant lots, the study designated mobile homes that had weathered hurricanes as blighted—with no documented evidence other than that the homes had a tendency to "blow over in storms."

High crime rates were also used as a justification to declare the area blighted, but there was no data to substantiate it. When the city was pressed to supply statistics, the information could not be retrieved purportedly because of glitches with a new computer system.

Buildings in good condition were declared "functionally obsolete." Generally, this means that any structure, if torn down, could be replaced with something that generated more tax revenue—not a particularly hard standard to meet.

Despite errors in the study, Mayor Brown remained committed to the redevelopment project. He said, "If we had a hundred homes and get two wrong, what does that mean?" He also inserted a racial element into the already volatile situation by saying that eminent domain had always been used to displace black people, but "Now, not all the faces are black. Now, all of a sudden it's tyranny!"

BEATING JEB TO THE PUNCH

Meanwhile, the grassroots campaign to reform the state's eminent domain law was a success. With the support of Governor Bush and Republicans and Democrats alike, the Florida legislature overwhelmingly passed a bill on May 4, 2006 (eleven months after the *Kelo* decision), which specified that the prevention or elimination of so-called blighted areas could no longer be considered as a public purpose.

The bill was hailed by property rights groups throughout the nation and considered a model for other states to follow. Florida had gone from having one of the most permissive eminent domain laws to having one of the strictest. Obviously, the new restrictions were not welcomed by the Riviera Beach City Council or by Viking Inlet.

Mayor Brown called a special emergency city council meeting six days after the legislature's vote. He intended to approve the development agreement with Viking before Governor Bush signed the bill.

According to a West Palm Beach attorney, Martha Babson sent the governor an e-mail alerting him on the eve of the meeting. The next day, however, the city passed the ordinance just hours before Governor Bush signed the bill. The criticism of the city's action was harsh.

Carol Saviak, the executive director of the Coalition for Property Rights, based in Orlando, said, "The city's actions are shameless and were taken with the clear intention to thwart the intent of the Florida Legislature."

Valerie Fernandez, a senior attorney at the Pacific Legal Foundation, a nonprofit legal advocacy group, said, "The city's backroom development deal is unlawful and it must be stopped."

Even Governor Bush weighed in and said, "I don't think [Riviera Beach] has a legal basis." The governor added that to rush to seize people's property "just seems unseemly to me."

A SHOT ACROSS THE BOW

In less than a month, three lawsuits were filed that alleged the city had not given proper notice of the meeting and that selecting Viking as the master developer was a violation of state law. The Cories and the Coalition for Property Rights were represented by the Pacific Legal Foundation. The Institute for Justice represented Princess Wells whose home and beauty salon business were threatened by condemnation. The law firm of Brigham, Moore also filed a similar suit on behalf of other owners.

In addition to the legal actions, the city had become embroiled in even more controversy. Two attorneys who represented the city's redevelopment agency resigned. It was also reported that the agency had recently spent nearly $1 million on consultants, but had nothing to show for it.

Controversy over the city's finances was not new. Two years earlier, a citizen's group on Singer Island had pushed for an audit of the CRA which showed that it was $7 million in debt. The report stated that the city was headed toward a "state of financial emergency." The property owner's group then asked the state to investigate.

Although spending seemed out of control, council members considered hiring Bernard Kinsey, a California consultant, to take over negotiations with Viking. The former Xerox executive and West Palm native had had experience working on a revitalization plan in Los Angeles.

The $3,000-per-day, six-month contract was approved, but not all CRA members were in favor of it. Councilwoman Ann Illes said, "We don't need [Kinsey] at that price. It's an insult to bring somebody in when we're so close to getting [the] deals done." But were the deals close to being done?

Due to the city's internal problems, bad publicity, and the lawsuits, Viking Inlet was forced to rethink its plans. Mike Clark, the president of its real estate arm, said, "Now I'm stuck with these properties but can't develop them because I can't fill in the puzzle pieces."

A SEA OF CHANGE

More bad news for the city and Viking Inlet came on November 7, 2006, when Florida voters approved a measure placing restrictions on the use of eminent domain in the state constitution. Then, in what could be considered the final blow for Viking's redevelopment plan, the city council voted 5 to 0 to approve a resolution to abide by the new eminent domain law. Mayor Brown, however, still vowed to challenge the state law in court.

The embattled mayor's plan to go to court and his support of other controversial projects had put him at odds with the predominantly white residents of Singer Island. Those residents backed a three-candidate slate for the upcoming election, including Rev. Thomas Masters, a black pastor who had lost to Mayor Brown in a previous election.

As a result, Masters soundly defeated Brown in a runoff election on March 27, 2006. After receiving a congratulatory telephone call from the Rev. Jesse Jackson in packed council chambers on April 4, the new mayor proclaimed that it was also the anniversary of Martin Luther King Jr.'s slaying and said, "The dreamer was killed, but the dream lives on."

Although the lawsuits against the city were dropped in May 2007, Viking Inlet's development dream seemed to vanish. Floyd Johnson, the executive director of the city's redevelopment agency said it was likely the project would be scaled back, and reiterated that no residents would be forced from their homes.

To date, Viking Inlet and its partners have spent more than $50 million to acquire property in the redevelopment zone. It is not immediately clear at the time of this writing what the company plans to do with the property. According to Johnson, however, talks with the developer are ongoing and cordial.

The citizens of Riviera Beach stand at a crossroads. If the past is any predictor of the future, the city's political bickering, racial tensions, backroom deals, and alleged financial improprieties will continue.

7

Story Snapshots

"THE DONALD" TRUMPED
(Atlantic City, New Jersey)

Donald Trump is not often humbled—especially by an elderly homeowner and two small businesses.

In the mid-1990s, the developer proposed to enlarge the Trump Plaza Hotel & Casino. Among other improvements, Trump proposed to build a parking lot for limousines across the street—an area occupied by several businesses and homes.

Vera Coking, who had lived in her home for more than thirty years, the family-run Sabatini's Restaurant which had occupied the same corner for more than twenty years, and Banin Gold Shop all refused to sell.

Trump, with the help of New Jersey's Casino Reinvestment Development Authority (CRDA), a state redevelopment agency, began condemnation proceedings in July 1994. The value assigned to the properties was based upon prices a parking lot would fetch.

The owners challenged the taking in Superior Court and argued that Trump's limousine parking lot was not for "public use." The court ruled that the parking lot could be used for public use, but evidence to the contrary showed that once the land was transferred to Trump, he could do whatever he wanted with the property—casino space, for example, would obviously be valued higher than a parking lot.

The court ruled against the developer. It said that the CRDA's condemnation amounted to giving Trump a "blank check" and ruled that the properties could not be seized.

James B. Perry, President of Trump Entertainment, would later say, "We should offer a reasonable price for those properties. But it's their property. They own it and have a right to say no."

FIVE STRIKES AND YOU'RE OUT
(Oklahoma City, Oklahoma)

Developer Moshe Tal submitted plans in 1995 to develop 1.4 acres of property he owned in downtown Oklahoma City into retail shops, restaurants, and entertainment facilities. The city, however, had a different use in mind.

Two years later, the city council said that it needed the land, for parks, recreational facilities, and parking, and condemned it. Appraisals for the property showed that it was worth $5 million. The city offered only $50,000—1 percent of its value.

Tal sued the city, but a trial court judge ruled that the city had the right to condemn property for "public use" under its powers of eminent domain. However, when the city announced plans to sell the property to a rival developer for $165,000, Tal went back to court and argued that the city had misrepresented the use of the land.

A trial court dismissed Tal's case. The Oklahoma Appeals Court and the State Supreme Court also ruled in favor of the city and against Tal's constitutional rights of just compensation and legal due process. He then appealed to the U.S. Supreme Court in 2002, but the court refused to hear the case.

Tal then filed charges under the Racketeer Influenced and Corrupt Organizations Act (RICO) in federal court against a rival developer that allegedly conspired with the city against him. The lawsuit was dismissed on June 29, 2006—after nine years of litigation and losing five times in court.

99 CENTS IS NOT ENOUGH
(Lancaster, California)

The City Council of Lancaster, California, gave an "800-pound gorilla" what it wanted—to crush its competition.

In 1998, the 99 Cents Only discount chain opened a store next to Costco in a large retail shopping center. Soon after 99 Cents started operation, Costco announced its intentions to expand and told the city it wanted the location where 99 Cents stood and requested that Lancaster use eminent domain to condemn it.

The city tried to convince Costco to expand in another direction, but the company insisted on the site. In fact, according to court testimony, the retailer not only threatened to relocate in a neighboring city, Palmdale, if it didn't get its way, but also vowed to close the store and leave it unoccupied.

Fearing the loss of tax revenue, the Lancaster redevelopment agency initiated condemnation proceedings. In an amazing display of financial incompetence, the agency proposed to buy the site where 99 Cents was located from its landlord for $3.8 million and then resell it to Costco—for only one dollar.

99 Cents sued the agency in district court seeking to stop the condemnation. From the start, the judge was skeptical of Lancaster's motives. Before the case went to trial, the city backed down and told the judge Costco had found another location in the city. Lancaster, however, refused to commit that it would not attempt to seize the 99 Cent location in the future. The trial went forward.

The judge ruled in favor of 99 Cents and said: "The evi-

dence is clear beyond dispute that Lancaster's condemnation efforts rest on nothing more than the desire to achieve
the naked transfer from one's private property to another."

When Lancaster appealed the decision and lost, its city
manager said the ruling was troubling and added that: "99
Cents produces less than $40,000 [a year] in sales taxes,
and Costco was producing more than $400,000. You tell
me which was more important."

PUTTING THE BRAKES ON EMINENT DOMAIN ABUSE
(Mesa, Arizona)

The City of Mesa, Arizona, thought that it was a good idea
to seize and demolish a brake shop and replace it with a
new Ace Hardware store. Randy Bailey, the owner of the
brake shop, didn't take too kindly to the plan.

Bailey's Brake Service has stood at a busy street corner since the early 1970s. Randy Bailey had purchased the
business from his father in 1995 in the hope that he, too,
would one day sell the shop to his own son. So when a local
businessman, without even approaching Randy, convinced
the city to condemn the shop, Randy fought back

The City of Mesa filed a lawsuit in Superior Court to
condemn Bailey's Brake Service. To make the deal even
sweeter for the developer, Mesa planned to pick up the
tab—using taxpayer dollars—for various fees that would
normally be incurred by the buyer.

More shocking than the city's willingness to take private
property for private use is the fact that four of the seven
city council members who voted to expand the redevelopment zone to include the shop had personal conflicts of interest. They, or members of their families, owned property
in the zone.

In October 2001, the Arizona Chapter of the Institute

for Justice countersued the city, asserting that Mesa was abusing its power of eminent domain. The trial court ruled for the city, but the Arizona Court of Appeals ruled unanimously in Bailey's favor. The court said that "the constitutional requirement of 'public use' is only satisfied when the public benefits" and that transferring private property to a developer to build a hardware store did not satisfy that standard.

Randy Bailey continues to sell brakes at the same corner that his father did.

THE NEW YORK TIMES: CORPORATE WELFARE RECIPIENT (New York, New York)

Three days after 9/11, a *New York Times* executive told officials and journalists, "We believe there could not be a greater contribution than to have the opportunity to start construction of the first major icon building in New York City after the tragic events of September 11." The fact that the company had been threatening to move out of Manhattan if they did not get a "sweetheart deal" to stay was, of course, not mentioned.

Already reeling from the economic consequences of the disaster, the city and state offered the *Times* a prime spot—one block from Times Square. The property was privately owned and consisted of eleven buildings and approximately thirty businesses.

Sidney Orbach, an owner of a building to be condemned, said a lot half the size of the one the *Times* wanted, and located in the same area, had recently sold for $180 per square foot. The deal the newspaper struck was expected to cost it only $60 per square foot.

In addition, the city and state signed a deal with the *Times* and developer Forest City Ratner Companies that included a ninety-nine-year lease totaling $85.6 million—well below market value, according to experts. Even though it is certain that the condemnation costs will exceed $85.6 million, the state and city will pay the difference, not the *Times*. But the giveaways don't stop there.

The proposed new skyscraper includes fifty-two floors and well over 1.3 million square feet of lease space, but only 900,000 square feet would be occupied by the *Times*. The rest of the space, chiefly the higher and more expensive floors, will make the newspaper a landlord of more than 400,000 square feet of prime Manhattan real estate.

The deal also includes $26 million in tax cuts for the *Times*. Adding up all the tax breaks and the anticipated subsidies to be granted, it is expected to cost taxpayers nearly $100 million.

No wonder a *Times* editorial titled "The Limits of Property Rights" cheered the *Kelo* decision and said it was a "welcome vindication of cities' ability to act in the public interest."

BEHIND THE STADIUM LAND GRAB
(Detroit, Michigan)

Freda Alibri agreed to sell the Detroit/Wayne County Stadium Authority a parking lot she owned as part of a plan to build two new sports stadiums. What she did not know was who was really behind the deal.

In the late 1990s, Detroit planned to build the Detroit Lions football and Tigers baseball stadiums next to each other. The Authority reached settlements that gave it title to all but twenty-four of the properties on the site and decided to use eminent domain to condemn the remaining properties—including the one-acre parking lot owned by Freda Alibri and her family.

Although the Alibris' property was across the street from both of the new stadiums, the stadium authority claimed it needed the Alibris' lot for parking. She sold the lot for nearly $270,000 based on the fact that it was to be used only for a parking lot and not transferred to a private party.

From the time the city chose the site for the two stadiums, Mike Ilitch, the founder of Little Caesar's Pizza and owner of both the Tigers and Detroit Red Wings, had, according to court testimony, been acquiring land across from the stadiums in the hopes of someday building a new hockey arena. However, it was discovered Ilitch had "loaned" the Authority the money to buy the lot (valued as a parking lot as opposed to a much higher value for a stadium) and intended to transfer title to him. Alibri sued to get her land back.

The trial court agreed with Alibri, but the appeals court overturned the lower court's decision and said that since the Alibris agreed to sell she had to honor the agreement. The Michigan Supreme Court, however, disagreed and ruled to return the property to Mrs. Alibri.

The pizza baron's scheme went up in smoke.

NOT A VERY BRIGHT SCHOOL DISTRICT
(Cumberland County, Virginia)

The Cumberland County School District made a blunder and wanted Mary Meeks and her family to pay for it. The end result: it cost taxpayers dearly.

In 2001 the district decided to sell an elementary school that had been boarded up for years and twenty acres it did not need anymore to Mary Meeks and her family for $110,000. The Meeks spent more than $400,000 on renovations, rented space, and opened the gym and auditorium for use by the community and senior citizens. Three years later, the district decided it wanted the property

back to reopen the building as a school—but offered only
$200,000.

The Meeks refused the offer. The district then deposited
$200,000 with a court and took the school by "Quick Take."
This process allows a county to take immediate possession
of the property, but in this case it also gave it the right to
collect the rent from the Meeks' tenants. The family sued.

Even though the Meeks' source of cash flow was cut off,
they could not use any of the $200,000 on deposit to pay
$30,000 in mortgage payments unless they waived their
right to appeal the eventual ruling of the trial court. The
Meeks stood their ground.

The case went to trial in February 2007. The jury
awarded the Meeks $850,000—more than four times
what the district offered. This was a serious blow to cash-
strapped Cumberland County with a population of only
9,017.

Part II

ZONING

8

CONTROLLING LAND LOCALLY

"[Government has come] to see themselves . . . as sculptors of neighborhoods, able to use the property owned by the city's people as the clay that [can] be molded into an ideal city."

—TIMOTHY SANDEFUR, *CORNERSTONE OF LIBERTY: PROPERTY RIGHTS IN THE 21ST CENTURY*

Government is legally required to compensate property owners in eminent domain cases. Through No-Compensation and Pay-To-Play Takings, however, it can take property without compensation and also *extort* land and money in return for certain approvals. It might be legal, but it's not right.

These takings, also known as "regulatory takings," can involve a city or county's control over the kinds of uses to which each individual property may be put and the physical development of that land. A means by which this is done is through local zoning laws or ordinances.

In addition to restricting land use, zoning policies can regulate dimensional requirements for lots, density (the number of homes that can be built per acre), space for hospitals, parks, schools, green space areas, and even places of historical significance.

Although there are basic land use categories such as residential, commercial, industrial, agricultural, and recreational, the actual policies vary from one city or county to another. One thing, however, is uniform: zoning issues (especially in urban areas) can become extremely contentious.

There has been an explosive migration of people and businesses moving from the cities to the suburbs. This movement, known as "urban sprawl," created legal and financial challenges for governments to construct highways, water, sewer, and other services to keep pace with rapid development.

Those people who escaped the cities for a better quality of life began to see subdivisions replacing green space. Retail and other commercial development followed in order to support the needs of residents.

Everyone wanted his piece of the "American Dream." Those who had achieved their dream, however, wanted to shut the door behind them and keep others from sharing in it.

Among the players are NIMBYs (Not In My Back Yard activists), BANANAS (Build Absolutely Nothing Anywhere Near Anything), and Greens, environmentalists with extreme no-growth philosophies. Often, one or more of these groups seek to form coalitions in an attempt to mask their true motivations.

For example, condo owners who do not want their views blocked would recruit Greens on some purported environmental issue. They would argue that their cause benefits the public, but in reality only a few condo owners would benefit.

LEGAL TUG-OF-WAR

There is no question that some parts of the country have experienced unbridled and poorly planned development.

Community leaders—often with little or no planning or development experience—are forced to institute local planning ordinances in an attempt to manage this growth. Some policies went too far; others not far enough. As a result, there have been an unprecedented number of lawsuits across the country over the past two or three decades.

One landmark Supreme Court case that demonstrated this new reality was *Agins v. City of Tiburon*. The case involved inverse condemnation which, broadly defined, is a legal avenue for owners whose property is overburdened with regulations. In these disputes, an owner sues the government in an attempt to force it to buy the property.

A property owner in the city of Tiburon, California, acquired five acres of land for residential development on a scenic hillside near San Francisco. Through local zoning ordinances, the city placed the property in a zoning classification that permitted only one single family home per acre.

The city justified its position by saying that "it was in the public interest to avoid unnecessary conversion of open space to strictly urban uses, thereby protecting against resultant adverse impacts such as air, noise and water pollution, traffic congestion, destruction of scenic beauty, disturbance of ecology and environment, hazards related to geology, fire and flood, and other demonstrated consequences of urban sprawl." Is that all?

The property owner sued and argued that the zoning was a violation of the Fifth and Fourteenth Amendments to the Constitution. After losing in state courts, the case was appealed to the U.S. Supreme Court.

The high court sided with the California Supreme Court. It ruled that—as long as zoning ordinances "advanced a legitimate government interest" and did not deny the owner the "economically viable use of his land"—the City of Tiburon's action did not result in a taking. A precedent was thus established: government is required to weigh private versus public interests when considering zoning ordinances. A very subjective test, to say the least.

Since most of the lawsuits in the 1980s cited infringement on a property owner's constitutional rights, the federal courts were flooded with cases. In a concerted effort to stop these cases from clogging the federal courts, the U.S. Supreme Court established what is known as the "ripeness doctrine," a major turning point in takings law.

In *Williamson County Regional Planning Commission v. Hamilton Bank*, a land use agency in Tennessee rejected a property owner's request to expand a housing subdivision. The owner filed an action in federal court in 1983 alleging a taking in violation of the Fifth Amendment.

Two years later, the high court ruled that the owner had not reached a final decision at the local level since it could have requested a zoning variance, or appeal, before the County Council. In addition, the justices said that the owner had to seek compensation in the state courts before coming to federal court.

This ruling gave government an unfair advantage since few property owners could afford the years of litigation it would take to go through this process. As a result, cities and counties, with nearly limitless legal budgets, now had the upper hand and could force property owners to bend to their will knowing that they could outspend owners in court.

PROCEED AT YOUR OWN RISK

As if the *Williamson* case were not bad enough, fast-forward to two later cases, *City of Chicago v. International College of Surgeons* (1997) and *San Remo Hotel v. City & County of San Francisco* (2005). The facts in each case are

interesting, but the opinions of the courts focus on procedure rather than on substance.

The conclusion reached in the *City of Chicago* case by
the U.S. Supreme Court was that government, at its whim,
could remove cases from state court to federal court. This
ruling baffled the legal establishment since nowhere in the
majority or dissent opinions was *Williamson* even mentioned. What happened to legal precedent?

In *San Remo*, the case was first filed in state court to
adhere to the ripeness precedent in *Williamson*. Property
owners lost in trial court, won on appeal, and lost a 4-to-3
decision before the California Supreme Court. When the
owners attempted to try their case in federal court, they
were told that the state court proceedings were "equiva-
lent" to a federal trial and the case was dismissed.

To summarize, in *Williamson* the court said owners had
to exhaust their efforts in state court before they could go
to federal court. In *City of Chicago*, the government can, ar-
bitrarily, move the case from state to federal court. In *San
Remo*, the court ruled whatever the state court says is good
enough for the federal court. Little wonder that property
owners filing takings claims feel as though they are step-
ping into a minefield!

A landmark case that involved the question of "temporary takings" wound its way to the U.S. Supreme Court in
1987. The case: *First English Evangelical Lutheran Church
v. County of Los Angeles*.

In 1978 a flood destroyed buildings on the church
property that served as a retreat for handicapped children.
The land was located in a canyon near a natural drainage
channel.

Los Angeles County then adopted a flood protection
plan which prohibited construction or reconstruction of
buildings. The church filed a lawsuit alleging that it had
been denied all use of its property.

State courts sided with the county, but the U.S. Supreme Court ruled that, under the Just Compensation Clause in the Constitution, the landowner was entitled to recover damages *before* it is finally determined that a regulation constitutes a taking. Temporary takings by government were now no different from permanent takings. This was a huge victory for private property rights advocates.

NO CLEAR STANDARDS

The no-growth community then came up with a new tactic in the form of building moratoria, or temporary halts to development. In the 2002 case, *Tahoe-Sierra Preservation Council v. Tahoe Regional Planning Agency*, a Lake Tahoe planning agency imposed a moratorium on the development of 700 single-family lots for a period of three years to study potential environmental impact on the lake from development. By what could only be described as land planning hijinks, the planning agency continued to wrangle over environmental studies and passed new ordinances and "temporary" moratoria that strung the property owners out for more than twenty years.

Finally, affected property owners filed suit in district court alleging that the actions were an unconstitutional taking. They prevailed, but lost on appeal before the U.S. Ninth Circuit Court of Appeals—a controversial court known for its liberal interpretation of the law.

The owners appealed to the U.S. Supreme Court. In a 6-to-3 vote, the high court sided with the Ninth Circuit. The majority of the justices agreed that the temporary taking in this case was different from the circumstances of the *First English* case in 1987 and that the property owners were *not* denied all economic use of their property.

This ruling was a clear setback for property rights advocates. Although it would depend on the circumstances of the case, government now had a legal tool to temporarily halt development without compensating owners. Ironically, the current Chief Justice of the Supreme Court John Roberts, whose nomination was hailed by many conservatives, represented the planning agency.

As can be seen from the foregoing cases, the courts still have not been able to develop a consistent formula to determine if a zoning action results in a taking. In addition, the confusion over ripeness further weakens the ability of property owners to defend their rights.

Pro-development coalitions, often referred to as the "Wise Use" movement, argue that the courts have done little to protect private property rights. Groups opposed to development, on the other hand, believe that the courts have not gone far enough. One thing they can both agree upon, however, is that rulings have done little to bring clarity to the situation.

Therefore, arrogant public officials, *using our tax dollars*, continue to resort to the tactic of "Sue us and we'll see you in court." Many owners cannot afford to roll the dice and are forced to accept unreasonable zoning conditions placed upon their properties. Public officials, NIMBYs, BANANAs, and Greens can then declare a victory for "the public good."

9

CITY OF PIRATES
(Pompano Beach, Florida)

"One of the greatest things in the American dream is to win against all odds."

—JIM STEPHANIS, FORMER CO-OWNER,
YARDARM RESTAURANT

The City of Pompano Beach, Florida, was bent on destroying the Stephanis brothers at any cost. For thirty-one years, it engaged in bureaucratic stonewalling and unending litigation to stop the restaurateurs from building a hotel. Why? Because the tower would block the ocean views of residents in a nearby high-rise condominium.

In the early 1970s, the seaside town on the east coast of Florida had become a haven for Midwestern retirees. Retailers were flourishing. Condominium towers—some as high as twenty-four stories—were being built. What was missing was a luxury hotel. Jim and Tom Stephanis saw the opportunity of a lifetime and seized it.

The brothers, self-made millionaires who got their start working in Chicago nightclubs, had the perfect site. Their highly successful Yardarm Restaurant sat on 1.3 acres on the Hillsboro Inlet and had a picturesque view of a historic lighthouse and the Atlantic Ocean.

Jim Stephanis confirmed with Pompano Beach city officials that the property's zoning had no restriction on building height or on the number of hotel rooms that could be built. He then hired an architect and plans were drawn up for an eighteen-story, 166-room hotel with a "world class" restaurant on the top floor. The estimated cost: $5 million. Since the brothers were flush with cash, they decided to build the hotel without borrowing a cent.

UNDERMINING THE PROJECT

When the plans of the Yardarm Inn were submitted to the city's building department in January 1973, the public outcry was immediate. Residents in the Hillsboro Light Towers, a fourteen-story condominium across the street from the site, were dead set against the hotel because it would block their views of the lighthouse and ocean. Neighboring single-family homeowners also joined the fray and protested that they didn't want Pompano to become "another Miami Beach." The battle had begun.

Over the next six months, the residents in the area pressured city officials to undercut the project by passing ordinances to limit the number of the hotel rooms and restrict the building height to ten stories. Even though the city's building chief, Walter Williams, went on the record and said there had been seventeen buildings built in Pompano that exceeded ten stories and recommended the brothers should be allowed to build up to eighteen stories, city commissioners imposed a 120-day building moratorium on the property.

The brothers fought back. They petitioned the city to exempt the property from the height limitation. According to later court testimony, Jim Stephanis claims that on the eve of the city commission vote, Mayor William Alsdorf, who was running for reelection, solicited him for a $10,000 bribe. Jim refused. Despite the mayor and vice-mayor voting against the exemption, the brothers won by a 3-to-2 margin.

The mayor, embittered from the defeat, suggested to angry residents that the city could use its powers of eminent domain to condemn the Yardarm property and make it a park. Instead, one month later, the residents filed a lawsuit against the city in circuit court to stop the issuance of a building permit. The suit was eventually dismissed.

In February 1974 the brothers submitted a full set of building plans to the city. Six days later, Fred Kleingartner, a city employee with the unofficial position as planning director, acting under the direction of Mayor Alsdorf and John Cartwright, the city manager, met with the Yardarm's architect to voice several objections. Plans were revised to address their concerns and later that afternoon Walter Williams, the chief building official, approved the building permit. The Stephanis brothers now had permission to start construction—so they thought.

MORE SHENANIGANS

Mayor Alsdorf, Cartwright, and Fletcher Riley, the chairman of the city planning board (and a member of a property owners association that opposed the project) struck back. They met with Williams and ordered him to revoke the permit.

A few days later, a pro-Yardarm commissioner was defeated in an election and the new commission voted 3 to 2 to reduce the number of the project's hotel rooms from 166 to 109. To add insult, the city manager wrote an official letter denying the Yardarm's building permit and returning the permit fees. The brothers had had enough and sued.

Yardarm v. City of Pompano Beach would take two years to reach trial. From the start, the city's lawyers sought delay after delay and made frivolous demands for information, which cost the Stephanis brothers additional time and legal fees.

In addition to decreasing the number of rooms allowed, the city commissioners passed a measure to increase the number of parking spaces required for the project. The brothers were forced to file another suit.

The judge ruled in favor of Yardarm in both suits, but the residents were not finished. The home owners association complained to the Federal Aviation Administration that the eighteen-story hotel would block the flight path of a nearby airport. The FAA was forced to investigate the matter and finally said that the tower posed no danger to aircraft.

On April 1, 1977, more than four years after they had submitted the hotel plan to the city, the brothers tore down the restaurant, and construction crews began sinking huge pilings to support the high-rise. Within two years, the Yardarm Inn would be a reality.

Unbeknownst to the brothers, the city had pulled a fast one. Another setback occurred relating to a discrepancy in the building permit. The original building permit had included permission to build docks—the new permit did not. To keep costs in line, the docks had to be built before the building because the site was too small to construct both at the same time. Since they thought the dock issue could be resolved in a timely manner, they halted construction. Big mistake.

INTENDED CONSEQUENCES

It took six months of negotiations with the city to resolve the issue. A judge oversaw the compromise and ordered the city to do everything necessary to get the docks approved through the Army Corps of Engineers. The Stephanis brothers had won another battle.

The brothers pushed on and resumed foundation construction of the hotel. They figured that by the time they received clearance from the Army Corps for the docks, they would still be able to build them without using a barge. But the city would not give up.

The judge had ordered the city to do everything it could to make sure the dock permit was approved by the Corps. Despite this, the city manager wrote a letter to the agency asking that the dock permit be denied because the docks would create a "navigational hazard."

Work was immediately halted and public hearings were scheduled. Then, Pompano revoked the Yardarm's building permit for a second time, citing a provision in the building code about lack of construction activity. Back to square one again—and into the courts.

By late 1977, due to several delays and runaway inflation, the project's construction costs were now estimated to be $12 million, an increase of $7 million over the past five years. Legal bills from fighting several appeals, all of which they had won, had also eaten away the brothers' cash reserves. They were forced to borrow money to keep the project going.

FULL STEAM AHEAD

The brothers hired a hotel consultant who advised them that the Yardarm site was too small to support the projected income and construction costs. He recommended Jim and Tom buy an additional 1.4 acres directly west, and across the highway, to enlarge the project.

This was a major decision for the brothers. Should they fold their cards and rebuild the restaurant or take a gamble and go forward with the hotel project—knowing full well that the city and neighbors would try to block them every step of the way.

Even though several city appeals and other actions were still making their way through the courts, the brothers were confident they would win the day. After all, they had won every suit and appeal to date.

In early 1981, with no building permit as yet in hand, the brothers bought a parcel of land for $500,000 and leased another. Six months later, it appeared the gamble had paid off. The brothers won another appeal and the city was forced to reissue the building permit.

Now that the Yardarm project was larger, the Stephanis brothers needed a $16 million loan. Although a newly elected mayor conceded that the brothers had won and said publicly that they should be allowed to build their project, some commissioners and the condo residents would not give up.

The city's lead attorney, Fort Lauderdale lawyer Henry Latimer, inundated the Stephanis brothers with a series of suits, motions, document requests, and depositions in order to discourage potential lenders from funding the project. The strategy worked.

Because of the specter of litigation, more than forty lenders refused to fund the proposed Yardarm Inn. The brothers were rapidly running out of money. They considered an investor's offer to buy the property for $8,250,000, but finally decided against it. Instead, they pledged the properties as collateral for a $3 million loan from a savings and loan association to stem their loss of capital.

The city struck again. In May 1985, Pompano yanked the building permit for the third time, citing that there had been no construction activity for four years. It was back to court again. The circumstances this time were much different. The brothers had a loan—and they would soon start running behind on their payments.

The brothers then learned, without being officially noti-
fied, that the city had voted 4 to 1 to rescind the Yardarm's
building height exception that had been granted *eleven years
earlier!* If the city took a second vote, the ordinance would
become law. Jim and Tom again raced back to court.

THE CITY GETS HAMMERED

The judge ordered the city not to vote on the measure, ruling that the evidence showed "that there has been a consistent
pattern by the city to frustrate the plans of the [brothers]." But
the ruling was a hollow victory. In March 1986, the brothers
had run out of money and were forced to file corporate bankruptcy. Their total indebtedness was $4.8 million.

The savings and loan foreclosed on the property and
put it up for auction. Only one bidder showed up at the
auction and paid $3.7 million for the properties—the City
of Pompano Beach.

In 1987 Yardarm sued the city and claimed that the
fifteen-year battle had amounted to an illegal forced "taking" of the property. It took four years for the case to reach
a non-jury trial.

During the trial, Pompano's attorney, Henry Latimer,
acknowledged there had been "wrongful acts" by Pompano
in the 1970s, but from 1981 to 1985 the brothers did not
have a valid building permit. He also tried to paint the picture that the brothers were blundering businessmen and
that lenders would not loan money on property that was
split by a highway.

The Yardarm attorney, Randy Adams, who was now
working on a contingency basis, argued that the primary
reason lenders shied away from the project was because
of the city's protracted and underhanded litigation tactics.
The judge agreed.

In a scathing opinion, the judge said that the city had committed "obstructionist and illegal acts" and condemned it for abusing the legal process with constant delays and appeals. He added that "the only thing promoted by the politicians was the protection of the property owners from having their view . . . obstructed."

The judge ordered a jury trial to determine how much the city would pay. Finally, after twenty years, the brothers felt vindicated. Yardarm asked for $35 million in damages.

The city appealed the ruling, and in April 1992 oral arguments were heard before a panel of three judges at the Fourth District Court of Appeals in West Palm Beach. The Stephanis brothers were confident they would prevail. They had, after all, endured and won nine lawsuits and several appeals.

ANTICS FROM THE BENCH

After eighteen months of waiting for a ruling, Rosemary Barkett, the Chief Justice of the Florida Supreme Court, wrote a letter to the Fourth District saying she was moving the case to the Fifth District Court in Daytona Beach. Jim, Tom, their lawyers, and much of the south Florida legal community were stunned by the ruling.

The maneuver to switch the case from one court to another was unprecedented. Even more shocking was that Chief Justice Barkett *chose* the three judges who would hear the case—rather than using the standard procedure to have judges randomly selected.

Adding fuel to the conspiracy theory fires, the circuit court judge who had presided over the original case, told attorneys Adams and Latimer that he had heard the Fourth District was ready to issue an opinion in favor of Yardarm just *before* the case was moved.

The Stephanis brothers were outraged. To pull a case from a court *after* the decision had been reached was beyond comprehension. Jim Stephanis was quoted in an interview by John Dorschner, a reporter for the *Miami Herald's Tropic* magazine, as saying, "I don't know how a woman feels when she's raped, but this was a lot like that. If this was the Wild West, we'd strap on our six guns and go get 'em."

Dorschner began his investigation and discovered a letter written by Judge John W. Dell, the Chief Judge of the Fourth District Court, to Justice Barkett. The letter stated that one of the three judges, Gary Farmer, had a conflict of interest and should be disqualified.

According to Dell's letter, Farmer, while in private practice, represented a developer by the name of Herman Corn whose case was still making its way through the courts. Dell stated that Farmer still had "a financial interest" in the case and that the Corn case and Yardarm cases were similar.

According to several attorneys, the cases were very different: Corn's suit was a civil rights claim in Federal court and the Yardarm's takings claim was in State court. In fact, both Adams and Latimer were well aware of Farmer's involvement in the Corn case and never objected to him ruling on the Yardarm case.

ATTEMPTS TO SETTLE

While the three new *hand-picked* judges considered the case, the two sides tried to negotiate a settlement. The brothers offered $16 million. The city countered with $7.5 million. The parties reached a stalemate after the brothers offered $13.5 million.

In September 1994, a year after Chief Justice Barkett

had played musical chairs with the case, the three judges on the Fifth District Court of Appeals released their opinion. Yardarm lost.

Despite a lower court judge saying that the city had abused the legal process with constant delays and frivolous appeals, the court cited three reasons why the Yardarm's takings claim was invalid: 1) because the brothers had turned down an offer from an investor to buy the property for $8.25 million; 2) that the brothers could have rebuilt the restaurant; and 3) that for four years in the early 1980s they had had a building permit and could build the project.

For the first time in their twenty-two-year saga, the Stephanis brothers had lost a case, but they persisted. They appealed the case to the Florida Supreme Court. The court refused to hear it. They then petitioned the U.S. Supreme Court. Six months later, the high court also declined to review the case.

A NEW LEGAL ANGLE

Yardarm attorney Randy Adams realized that he needed to come up with a different approach and more legal firepower. He provided documents to Margaret Cooper, an attorney with the West Palm Beach firm of Jones, Foster, Johnson & Stubbs. After careful review, she and her firm decided to take the case on contingency—even though the firm had never done so in the past.

Their strategy was to reinstate the original lawsuit by arguing that the brothers' civil rights had been violated. In addition, they would contend that the city violated the due process clause of the U.S. Constitution: no state can "deprive any person of life, liberty, or property, without due process of law; nor deny to any person within its jurisdiction the equal protection of the laws." Their plan worked.

In August 1996, Broward County Circuit Judge W. Herbert Moriarty ruled that the *original* Yardarm case could be reinstated and set it for a non-jury trial. And once again, Pompano's attorney, Henry Latimer, and his team went into high gear with a barrage of motions intended to stall the case. They knew the Yardarm attorneys were taking the case on contingency and hoped to force a settlement.

The case reached trial. Judge Moriarty ruled that Pompano was liable for monetary damages and took the opportunity to reprimand the city. He wrote: "No court could adequately address the harm which has befallen Yardarm during the litigation process." He cited that the city "abused the litigation process it is theoretically designed to protect and deliberately disobeyed [court] orders requiring Yardarm" to endure lengthy, protracted, and expensive litigation which forced it into bankruptcy and foreclosure.

VICTORY AT LAST?

The jury's verdict took only a few hours. They ruled in favor of Yardarm and awarded $19.2 million in damages! The judge also ruled that Yardarm was entitled to $5,241 in daily interest that would accumulate until an appeal to a higher court reversed the decision or the city paid the judgment.

Clearly, the city had its back against the wall. The possibility that the City of Pompano itself would have to file bankruptcy to avoid paying the judgment was real.

Pompano attorney Latimer immediately filed an appeal to the Fourth District Court of Appeals and then continued his stonewalling tactics by filing several motions to disqualify judges in the Fourth District Court of Appeals. Some court watchers speculated that he was trying to whittle down the selection to three judges who might look favorably upon the city's legal position and reduce the award to Yardarm.

Yardarm then made the last of several offers to settle: $19,550,000 to be paid in yearly installments of $4,500,000 including 8 percent interest. Two weeks later, the city countered with an offer for $10 million payable $1,000,000 per year for ten years and with no interest. The offer was flatly rejected by Yardarm.

On October 9, 2002, the Fourth District Court of Appeals reversed the lower court's ruling. Yardarm suffered its second, and most crushing, defeat.

The majority opinion, written by Judge Carole Y. Taylor, raised eyebrows throughout the south Florida legal community. She wrote that, despite all the evidence of the city's delaying tactics, "Yardarm was not deprived of substantially all use of its property" and, therefore, there was no taking of private property, nor were Yardarm's civil rights violated.

The most astonishing part of Taylor's opinion was that the case "is devoid of any evidence that adverse actions taken by the administrative officials, such as the Building Official, City Manager, and other department heads and employees, were directed or required by the City Commission." The facts of the case *clearly* showed otherwise.

The Stephanis brothers and their lawyers pressed on. Two high-powered Los Angeles appellate lawyers, Michael Berger and Gideon Kanner, who had argued some landmark takings cases before the U.S. Supreme Court, joined the team.

An appeal was filed with the Florida Supreme Court, but it refused to review the case—for the second time. An appeal was filed with the U.S. Supreme Court and it, too, declined to review the case. Jim and Tom Stephanis' thirty-one-year quest for justice had come to an end.

Within a year after the U.S. Supreme Court turned down

the case, Jim Stephanis, 73, suffered a massive stroke. He has made a miraculous recovery and works as a wine manager at a Crown Liquor store in Boca Raton. Tom, 80, is in quiet retirement with his wife in nearby Fort Lauderdale. And the city?

Although the city's actions to revoke the Stephanis brothers' building permits contributed to their downfall, the wisdom of various business decisions they made is still open to question. What is *not* open to question is that the City of Pompano Beach spent $3.7 million for a park and, reportedly, $3.5 to $5 million in legal fees for more than thirty-one years—all so that the precious ocean views of a small group of residents could be preserved.

10

BUZZARDS CIRCLING OVER BUZZARDS BAY
(Dartmouth, Massachusetts)

"Private property rights have never been allowed to take precedence over our shared national values and the preservation of our country's heritage."
—EMILY WADHAMS, THE NATIONAL TRUST FOR HISTORIC PRESERVATION

What happened to a Massachusetts family at the hands of preservationists should never happen in America. The story, however, serves as a prime example of how far some supposed "do-gooders" will go for the sake of preservation.

Nestled at the head of the Slocum River in the southern part of the state is the small, historic town of Russell Mills, an early 1660s settlement that eventually became the town of Dartmouth. The community has more than sixty miles of Atlantic coastline, mature forests with hiking trails, and farmland that produces corn, alfalfa, and pastures for livestock.

In 1969 Joseph and Maria Hill bought a dilapidated dairy farm from a development company to foster a work ethic in their three sons and two daughters. The Hills' property, along with surrounding properties, was then rezoned from agricultural to residential in the mid-1970s.

The rezoning permitted one home to be built on every acre, which, of course, made the land prime development property. Several developers sought to buy land in the area, but the Hills refused to sell. In fact, since they wanted to preserve the area's natural beauty, the family bought property themselves and increased their holdings to more than 1,110 acres—part of which fronted on a pristine saltwater cove known as Buzzards Bay.

A WIN-WIN SITUATION

The family's dairy farm consisted of a rebuilt barn and three grain silos to store feed for a herd of 250 cows that produced 6,000 pounds of milk per day. Income from the operations supported the family and five employees. The barn's milking parlor, however, needed to be modernized. In addition, a watering pond for the cows needed to be enlarged.

The Hills obtained a $70,000 quote from an equipment dealer to install a computerized system that would not only make the milking operation more efficient but also enable them to better monitor the health of the livestock. Although they had only a small debt of $300,000 on the property, they looked for ways to raise the capital without borrowing.

The Hills learned that the town was in need of gravel to cover a landfill. By enlarging their pond they could provide the material and at the same time raise the money for the milking equipment.

The cost of gravel sold to the town had steadily risen each year, but the Hills submitted a bid that was lower than any the town had received in the past three years. The response from local residents caught them completely off guard.

A former town planner led the opposition against the Hills' bid by stating at a town meeting that the new gravel operation would endanger children—although there were very few children living in the area. Interestingly, the previous owner of the Hills' property had sold gravel to the town to repair a road and no such concerns were ever raised.

The town's building inspector and fire chief, who lived near the farm, wrote letters to the city council opposing the bid. Other residents joined in and falsely alleged that the Hills intended to develop their entire property which would, in turn, spoil the rural lifestyle that many people who had moved from the city wanted to preserve.

Eventually, the gravel bid was awarded to the highest of six bidders. The lack of a rational basis for the award and the hostile treatment the family had endured was a precursor of things to come.

Over the course of several years, Maria's husband, Joseph, had suffered several heart attacks. With one son enrolled at the dairy science program at Virginia Tech, the Hills family had to find a way to raise enough money to keep the farm operating until he graduated.

Maria was aware of a state program to buy property rights in order to preserve property from development. Looking to raise the needed capital, in April 1981, she met with the Massachusetts agricultural commissioner to inquire if the state would be interested in buying the underlying development rights to the land. Maria was told that the agency was very interested but that the legislature had not provided it enough funding. However, the commissioner referred her to representatives of private investors who pooled funds together to preserve property. One such person was Davis Cherington.

Maria met with Cherington and made it clear she was not interested in selling the land—only the rights so that the property could be preserved. The value of those rights would be the difference between the current market value of the land if it was developed into one home per acre less the value of the farm property.

Eventually, a trust Cherington represented made an "offer." The group proposed an option to purchase the property—despite Maria's insistence she would not sell.

The group's plan was to utilize a combination of private donations with state and local grants, subdivide the choicest part of the land into estate-sized lots, and then sell the remaining parcels to groups such as the Audubon Society. One catch: this would only be accomplished when, and if, they could raise the money. The trust even went so far as to say that the farm operation could remain but that the Hills would have to move off the land.

The Hills, of course, turned down the proposal. They also rejected schemes from the Massachusetts Audubon Society and other preservation trusts that would simply remove the property from the market without providing a firm commitment on how and when to pay for it.

At one point, realizing that their tactics would not work, the president of a local trust, known as the Slocum River Trust, and a member of the Audubon Society, contacted the Hills' bank and inquired about buying the Hills' mortgage debt. By owning the debt, they figured that they could eventually force the Hills to sell or even foreclose on the farm. The bank told them to deal directly with the Hills.

"FRIENDS" OF THE COMMUNITY

In order to restart the dairy operations after their son graduated, the Hills decided to sell the herd and lease cropland to a farm family in a nearby town. They figured the income would provide them enough money to sustain themselves and pay their annual property taxes of $35,000.

In 1985 a developer who was building a subdivision in the area approached the Hills. He asked them if they would sell 350 acres to him before the zoning was changed from one home per acre to one home on every two acres. The Hills were shocked.

They then learned that a group of residents calling themselves "Friends of Russell Mills" (FORM) had been quietly circulating a petition to conduct a special town meeting to rezone the property. The rezoning would, in theory, devalue the land by half. Even though the family had no intention of selling or developing the land, they were forced to protect their interests.

Maria Hill hired a civil engineer to prepare a preliminary subdivision plan. This document would legally preserve the right to build one home per acre.

On the afternoon of January 14, 1986, the plan was filed. At a town meeting—held at 11:00 p.m.—the town planners voted against the application and then changed the zoning to two acres per home. The denial, according to the Hills' attorney, was a first—not only for the town but also in state history.

Maria was furious. She hired a lawyer and filed a lawsuit in the Land Court of the Commonwealth against the town. The suit asserted that the town's planning board had bowed "to the special interest pressure of FORM," and violated both the family's state and federal constitutional rights by unlawfully denying their plan. Then things got really nasty.

According to Maria, before the first court hearing, FORM's attorney threatened her on the courthouse steps by saying that the group, if necessary, would litigate the case all the way to the U.S. Supreme Court. In addition, unless the family sold FORM the property, it would discourage other potential buyers, make certain banks would not loan them money, and then force the Hills to lose their land through foreclosure.

THE CONSPIRACY THICKENS

On April 6, 1989, the Land Court judge decided in favor of the Hills. The judge ruled that the town's planning board had exceeded its authority. He also said that since the Hills filed their plan before the property was "down-zoned," they were entitled to the one-home-per-acre zoning designation.

Soon thereafter, FORM, which appeared to be closely related to the group that had originally petitioned for the two-acre zone change, appealed the decision. The court dismissed the suit. The judge said the group did not have "standing"—the legal right to appeal the ruling—since it had been the town of Russell Mills that had made the motion to change the zoning.

Maria contents that town officials and FORM were conspiring to prevent her and her family from legally developing their land if they wished. For example, it was later discovered in court proceedings that the town's legal counsel wrote to FORM's attorney stating that he had "cautioned" the building commissioner's office to inform anyone wanting to build on the Hills' property "to proceed at their own risk" since appeals were anticipated.

A regular meeting of the Dartmouth Planning Board was then held in late 1989. Since FORM could not appeal the ruling, it offered to pay the town's costs to continue appeals. In a 3-to-2 vote, the Board agreed to file another appeal "provided no public expenses are incurred."

The legality of the move was highly questionable. FORM, it was agreed, would simply make donations to the town's "gift fund" to cover the town's costs—a feeble attempt to mask the true motivation. It appeared that the group's tactic was to force the Hills to spend hundreds of thousands of dollars on attorneys.

In 1990, as the appeals began, Maria hired an appraiser

to determine both the market value of the land and also its distress sale value—a reasonable price if the land were to be sold quickly. The fair market value of the property was determined to be $20,000 per acre, including the value of the home and farming improvements, for a total of $29.5 million. The distress sale value was estimated to be $17.7 million.

The prospect of prolonged litigation motivated the Hills to sell. Another contributing factor was the mysterious burning down of an old barn and a house on the property.

In late 1991, the Hills entered into negotiations with a British company and agreed to exchange their property, valued at $16.5 million, for one that the company owned in Germany. It appeared their problems were solved. However, when the townspeople learned of the deal, they openly stated that they would challenge any permit the company sought in court.

At the last minute, negotiations broke down with the British company and the deal fell through. The Hills were devastated.

ANOTHER BUYER TO SAVE THE DAY

The family was becoming increasingly desperate at this point. Over the past seven years, they had been forced to remortgage their property twice to pay for legal fees, engineering costs, and to make mortgage payments. Their debt had now ballooned from $300,000 to $4 million.

While reading the *Wall Street Journal* in July 1992, Maria noticed an advertisement for "a property exchange." She replied to the ad by writing a letter to an attorney in Zurich, Switzerland. Soon after, the attorney sent his agent, Josef Huber, to inspect the property.

For the next few months, Maria shared information

with Huber, and in October they came to a verbal agreement for a joint venture (a partnership) rather than an exchange since Huber claimed that an exchange would take too long. The Hills agreed that Huber would satisfy all the bank debt and pay them roughly $10,000 per acre for the land—around $11 million.

The president of the Hills' mortgage company, who had been very supportive over the years, retired. His replacement was under pressure from government bank regulators to liquidate all mortgages that were secured by land.

The bank issued a notice of foreclosure and Maria contacted Huber. In an effort to delay the foreclosure, Huber produced some indication of financial backing from his client. The letter, however, was vaguely worded and did not satisfy the bank.

Huber then claimed that his backers needed more information to complete the transaction. Maria gave him engineering plans, projections to build out the project, and other documentation while Huber continued to assure the Hills that he was moving ahead.

TIME TO FIGHT BACK

The town of Russell Mills, with the financial backing of FORM, had appealed the legality of the Hills' subdivision plan three times. In all three cases the court ruled in favor of the Hills. Maria had had enough and decided to sue the group.

In September 1994, the Hills' company, Moorhead Corporation, brought a civil rights suit in Bristol Superior Court against FORM, its members, and other related land trusts. The Moorhead suit claimed that FORM had committed interference with advantageous business relations, abuse of process, and civil conspiracy in preventing Moorhead from lawfully developing their property.

FORM and the other defendants sought to have the lawsuit dismissed by describing it as a suit known as SLAPP (Strategic Litigation Against Public Participation). This type of lawsuit alleges that the right to free speech has been violated.

According to Massachusetts law, dismissing a SLAPP suit is quite simple. However, the court found Moorhead's claim was valid and ruled against the dismissal. FORM, and its members, could now be personally libel for millions of dollars.

The bank was close to foreclosing. In late 1992, Huber had said that the money to pay off the bank debt was forthcoming. However, in order to protect the property from foreclosure, the Hills placed it in Chapter 11 bankruptcy. This filing—which is overseen by a court-appointed trustee—is used to delay foreclosure proceedings and give the debtor more time to reorganize his business affairs and pay his debt.

Huber, and a corporation in which he reportedly had an interest, did come through. However, instead of paying off the mortgage, Huber bought the mortgage from the bank at a discount, with the intent of foreclosing on the Hills and owning the land outright.

LAST DITCH EFFORT

For six months, Maria Hill tried to negotiate a settlement with Huber, to no avail. By July 1993 the Hills were forced to borrow from friends to file a lawsuit against Huber and the corporation he represented.

Litigation, which included a series of bankruptcy hearings, continued over the next three years. Maria and Joseph, both in their early seventies, realized that the prospect of losing the land they had owned for more than thirty years was real.

The civil rights suit they had filed against FORM and its members had become an "asset." When the judge ordered the trustee to sell all assets to satisfy debts, anyone who owned the lawsuit could proceed with the claims against FORM.

FORM and its members saw an opportunity to avoid being liable for millions. It offered the trustee $1,000 dollars for the Hills' lawsuit. The Hills, even though they could not even afford to heat their home, bid $1,500. FORM increased their bid to $2,000—along with a written threat to sue the trustee for their expenses if he awarded it at a higher amount to the Hills.

Knowing that federal law called for awarding the asset to the highest bidder (even if by one dollar), Maria bid $2,250. According to Maria however, the trustee awarded the lawsuit to FORM for $2,000.

Maria filed an appeal with the appeals courts and lost. She then appealed to a circuit court but lost again. The property was listed for sale at auction.

The Hills attempted to prevent the sale through some bankruptcy court maneuverings, but in January 1996 the property was sold to Indurama Finance USA, a company Maria believed was controlled by Huber, for only $5.5 million. An appraisal, based on land sales in area at the time, listed the market value of property at $12.5 million.

On September 3, 1997, the Hills—including Maria's 97-year-old mother who lived in a caretaker house on the property—were evicted. Joseph, however, would not give up the fight. Without money for lawyers, he began writing the U.S. Supreme Court requesting that it hear their case.

One year later, however, while writing a legal brief in response to a motion filed by the town, Joseph died of a heart attack. Soon thereafter, the high court refused to hear the case.

EVERYONE WINS BUT THE HILLS

In 1999 an organization known as the Trustees of Reservations and the Dartmouth Natural Resources Trust formed a partnership to buy the 1,100 acres formerly owned by the Hills. According to their newsletter, the group signed a purchase and sale agreement with Indurama, which since foreclosing on the property, "has been looking for a suitable buyer to purchase the lands."

The Hills' property was sold off in pieces. Since sources of funds to purchase the property came from a combination of private donations, federal and state grants, and sales of parcels, it is difficult to determine how much was paid. One thing is a matter of record: Indurama sold 757 acres for $5.5 million and recouped its initial investment. More than likely, they pocketed a substantial profit for the remaining acreage.

It appears another winner was Neil Van Sloun, a local resident and adversary of the Hill family for many years. Van Sloun purchased portions of the property to operate a commercial nursery.

Maria Hill, now eighty-four years old and in failing health, subsists on her Social Security and lives with her son in New Bedford. Her only hope at this stage in life is that the telling of her family's story may help other property owners who become targets of preservationists.

11

HIGHWAY ROBBERY
(Rochester, Minnesota)

"What better way for government to acquire land cheaply than to condemn it after its development potential is destroyed by regulatory conditions?"
— FRANK KOTTSCHADE, MINNESOTA HOMEBUILDER

Cities often impose conditions on property that force owners to donate land or build public improvements in return for development approvals. These requirements are known as exactions—a politically correct term for extortion.

Franklin Kottschade (pronounced "cot-shoddie") is an American success story. He was born and raised on a farm forty miles north of Rochester, Minnesota (home of the renowned Mayo Clinic) and attended a one-room schoolhouse for eight years before high school.

Frank worked on highway construction jobs to pay for college, where he met his wife, Bonnie. He founded North American Realty, Inc., in 1972, a small homebuilding and real estate brokerage company. Over the years, North American has built more than 1,400 homes for Minnesotans across the economic spectrum—ranging from single-family home subdivisions, rental apartments, senior housing, mobile homes, to town houses.

Frank has been an active member of the National Association of Homebuilders (NAHB) and has always had a strong commitment to give back to the community. He served as the volunteer chairman of the Olmsted Facilities Commission, which was responsible for the development of the new joint City of Rochester/Olmsted County government center. He oversaw every aspect of the four-and-a-half-year project, from site planning through final construction—including the acquisition of forty-five separate parcels of land at market value prices, without resorting to litigation.

Frank also served as chairman of the Design and Construction Committee for the Rochester School District. The position entailed the management of public improvement projects and the renovation of schools—an experience that provided him with firsthand knowledge and respect for the challenges faced by local governments in managing growth.

Frank's community involvement helped to cement relationships with public officials and residents alike. Despite all the apparent goodwill, however, none of it seemed to matter after he announced plans to develop a 220-acre parcel in 1992.

HITTING A BRICK WALL

Before Frank closed the transaction, he met with the Rochester planning staff to discuss plans to build a shopping center on the eastern side of the property. He claims the department was concerned that the premature announcement of the shopping center would trigger land speculation in the area around the development. As a result, he put his project on hold for two and a half years.

In 1994 Frank submitted an application to rezone the eastern portion of the property for a proposed shopping center. The planning staff, according to Frank, claimed that the roads were inadequate and could not handle the volume of traffic created by the project. His request was denied.

At the same time, Frank submitted a plan to construct 104 town houses on 16.4 acres of the property that was already zoned for that use. The project would have fulfilled a need to provide more affordable housing in the area.

The Planning Department recommended approval of the project but attached nine excessive conditions that would drastically reduce the number of town houses and render the project uneconomic.

One demand was that Frank build a man-made lake, which the staff believed would enhance the aesthetics of the project. Building a lake was not the most environmentally prudent move; it would require digging up a low area on the property where a creek flowed.

Frank agreed to the city's request, but the State Department of Natural Resources objected to the plan and said it did not want the creek to flow into the lake. The city's demands for the lake and the state's refusal to allow construction of it put Frank "between a rock and a hard place."

A COST OF DOING BUSINESS?

The planning staff came up with another condition. They proposed that Frank build an entirely new frontage road along a major highway to accommodate traffic generated by the planned town houses. This requirement, he believed, was unnecessary and excessive. Frank commissioned a traffic study which proved his point that the road was not needed. In testimony before Congress (and on behalf of the National Association of Home Builders), Frank would later state that the city disregarded the report and produced no evidence to refute its conclusions.

In addition to the demand to build the frontage road in exchange for its approval, Frank testified that the city required him to *donate* a fifty-foot strip of his land in the event the city *might* need it to widen a road in the future. Frank objected, particularly since the city did not base its demand on any traffic study. He did, however, agree to wait until the city had completed its study of the road-widening but eventually decided to shelve the project for six years due to market conditions and other business-related reasons.

In early 2000, Frank requested information regarding the proposed roadway expansion. According to his Congressional testimony, the planning staff had not conducted any studies and simply insisted that the fifty-foot strip of was necessary.

Although Frank's studies showed his land was not needed for the road, he offered to yield a forty-one-foot-wide strip in order to move the project forward. He testified that the city demanded all fifty feet—with no justification.

Rochester's extortion tactics did not end there. It required that Frank construct ponds for the purpose of channeling storm water runoff from surrounding properties. The plan was to drain 3,000 acres of nearby lakes slated for development. The city then intended to charge *other* developers about $2,000 per acre for digging the ponds. The city, according to his testimony, never intended to reimburse Frank for the construction.

In May 2000 the Planning and Zoning Board recommended approval of the town house project even though Frank clearly demonstrated to them that the exactions made his project uneconomical. The city's demand that he donate fifty feet of property and construct drainage ponds reduced the number of town houses he could build from 104 to 26.

The construction of the ponds and the related work would have increased the development costs for each town house by more than 300 percent. This increase in cost would, therefore, push the price of the units far beyond the reach of the targeted middle-class buyers and much higher than market prices in the area.

PULLING THE TRIGGER

The next step in the approval process was to go before the Rochester City Council. The council approved the project with all the conditions imposed by the planning board. Frank sought relief from the requirements and applied for variances, or exceptions, to the exactions. After months of review by the city council and zoning appeals board, the city remained firm in its position.

Nine years after he had purchased the property, the project was at a stalemate. Frank firmly believed that his constitutional rights had been violated. Even though he subscribed to the belief of many developers that "Your first lawsuit is your last permit," he felt forced to sue the city.

After discussions with his legal counsel, Frank decided to file a suit in federal court—not state court. His lawyers knew that if the state court ruled against him, Frank would not be permitted to pursue the claims in federal court. In addition, the state proceedings would have made the litigation more lengthy and expensive.

The city responded with a motion to dismiss the case using a 1985 Supreme Court case, *Williamson County v. Hamilton Bank*. Based upon its legal precedent, the city focused on two aspects: 1) that Frank failed to get a *final* decision from the City; and 2) and that he failed to sue them in state court.

In January 2002, the court agreed with the city's position and dismissed the case. It said that Frank's claims were not "ripe" and that he should have sought relief in state court before pursuing his claims in federal court—even though he claimed that his *federal* constitutional rights were violated.

A JUDICIAL CONUNDRUM

A month later, Frank filed an appeal with the Eighth Circuit Court of Appeals. In his suit he utilized a 1997 Supreme Court ruling, *City of Chicago v. International College of Surgeons*, which essentially modified the *Williamson County* decision by allowing the *government* to move land-use cases like Frank's from state court to federal court. The argument: If the government can move a federal takings case to federal court, why can't a private property owner? The court was not sympathetic.

Although the appeals court acknowledged in its written opinion that Frank's legal position "has the virtue of logic and is tempting," it sided with the city and agreed the case should be dismissed. It also said that if Frank lost his case on the state level, he could then file for an appeal at the U.S. Supreme Court.

However, one judge said during the hearing that getting the high court to hear the case would be like winning the lottery. He was right. In October 2003, the U.S. Supreme Court declined to accept Frank's case.

Meanwhile, Frank's battle on the local level was far from over. While he was waiting to see if the U.S. Supreme Court would accept his case, the Minnesota Department of Transportation (Mn/DOT), using its powers of eminent domain, filed papers to condemn a portion of Frank's property to relocate a highway. Not only did the state condemn a large portion of the 16.4-acre parcel he was attempting to use to build the town houses, they sought to take a total of 28 acres.

In theory, Frank was not opposed to the government's use of eminent domain for public projects—as long as owners received just compensation. But in Frank's case, the state's offer for roughly ten cents on the dollar was not acceptable.

Mn/DOT valued the land at $875,000. Frank and his experts believed the property was worth $8 million. When Frank asked a state official to explain the basis for the low-ball figure, the official told him the value of the land had been depressed—due to the onerous development conditions that the city had placed on the property!

The state took title to Frank's land. He vowed to fight on and be heard before a court-appointed commissioner who would listen to testimony on the valuation from all sides.

BEYOND PERVERSE

Just when Frank thought things could not get any worse, they did. If (at some point) he would be able to develop his property, the city would require him to enter into a development agreement. These types of agreements set forth the terms and conditions under which a property owner could develop land, and only become final after they are authorized by the city council and executed by the mayor and city clerk.

Although the agreement at this stage was in draft form, it stated that a portion of the money Frank would receive from the Mn/DOT must be paid to the city. Why? It had required him to donate that land to the city in the first place. In other words, the city wanted Frank to pay *it* for land that the Mn/DOT took from him!

Meanwhile, as part of a Mn/DOT road project, the City of Rochester imposed a property tax assessment on Frank's property for more than $1.7 million. It claimed since the property it would seize from him would provide better traffic access to the site, the land would be worth more and therefore he would have to pay higher taxes.

Clearly, this is one of the more appalling twists in Frank's horror story. The city now contends that his land has increased in value. However, the Mn/DOT, during an eminent

domain proceeding, claims his land has fallen in value because of the excessive conditions the city has placed on it.
Two different levels of government, therefore, have wielded
their power and played off each other to put the squeeze on
Frank from putting his property to a reasonable use.

Frank sued Mn/DOT in Olmstead District Court over
losing driveway accesses into the property. After a jury trial
in April 2007, he was awarded a $1.5 million judgment
against the state.

Frank then testified about his fifteen-year ordeal before
a Congressional Committee in support of the Private Property Rights Implementation Act. Passage of the bill would
open the federal courthouse doors to land-use disputes like
Frank's.

The measure passed the House of Representatives by a
231-to-181 vote with the support of 88 percent of the Republicans and 81 percent of the Democrats. Despite this
overwhelming bipartisan support in the House, the bill
failed to even reach a vote in the Republican-controlled
Senate.

Perhaps if the senators were personally subjected to the
sort of extortion tactics Frank endured, they would sing a
different tune.

12

STORY SNAPSHOTS

PAY RANSOM OR ELSE
(San Francisco, California)

Through the City of San Franscisco's Hotel Conversion Ordinance, 500 small hotels were forced to set aside a percentage of their rooms at below-market rates for the homeless and other low-income residents. In order to avoid the regulation, owners were forced to pay a fee. In the case of Tom and Robert Field, owners of the San Remo Hotel, the fee was $567,000.

The Fields, who had acquired the 97-year-old building in 1971 and had spent more than $1 million to renovate it, felt that they should not have to foot the bill for social burdens that all taxpayers should bear. They filed a lawsuit in state court claiming the new ordinance revoked their zoning rights and that the regulation was a violation of the Fifth Amendment of the Constitution.

The Fields lost the case. They were forced to pay the city in order to rent the restricted rooms to tourists. They then decided to appeal their case to get their money back. The state appeals court ruled in their favor, saying that the San Remo and other hotels like it should not be forced to subsidize housing for the homeless.

During the oral argument, Justice Janice Rogers Brown told the attorney representing the city: "You are taking the property owner's right to use his property as he wants and you are saying he is able to continue that use if he pays you ransom." That, she continued, would be like her taking the lawyer's car and refusing to return it unless he met her demands. "In many jurisdictions," Brown said sharply, "that would be theft."

The city appealed to the California Supreme Court, which overturned the lower court's decision. The Fields appealed to the Ninth Circuit Court of Appeals, a court known for its liberal leanings, and lost.

The Fields then appealed to the U.S Supreme Court. The high court ruled 5 to 4 that federal courts should not be the forum for takings cases.

The brothers endured twelve years of litigation only to have the federal courthouse door slammed in their faces.

A TWENTY-YEAR PROPERTY FREEZE (Lake Tahoe)

Lake Tahoe is known for its extraordinarily clear water and beautiful mountain vistas. The Tahoe Regional Planning Agency (TRPA) would go to great lengths to keep it that way.

To maintain the lake's clarity and to control development, the states of California and Nevada formed the TRPA to "coordinate and regulate development . . . and conserve its natural resources." In order to achieve its goals, the TRPA moved in 1981 to stop the proposed construction of new condominiums, apartments, and subdivisions for a period of thirty-two months. However, due to a series of temporary moratoria and wrangling over environmental studies, this temporary freeze lasted for more than twenty years.

In protest, approximately 2,400 property owners around the lake formed the Tahoe Sierra Preservation Council

(TSPC). At the time of their property purchases, these owners had had the right to construct homes.

The TRPA sued in District Court and argued that the moratoria constituted a taking under the Fifth Amendment. It won, but the Ninth Circuit Court of Appeals reversed the ruling. On appeal in 2002, the U.S Supreme Court ruled that the imposition of a moratorium does not constitute a taking requiring any compensation.

Because of the implications of the ruling, property owners like Dorothy Cook, who paid $5,500 for a lot twenty years earlier, cannot build a retirement home. She did, however, have the privilege of paying property taxes for twenty years for an asset that has little or no value.

IT'S OKAY TO PRAY, BUT NOT TOO OFTEN (Denver, Colorado)

The First Amendment states that the government shall not make any law that prohibits the free exercise of religion. Apparently, a certain city official believed that a Denver zoning ordinance superseded the Constitution.

Every week, about a dozen women would gather at Diane Reiter's Denver home for Bible study, prayer, and dinner. In October 1998, a city zoning administrator, citing a municipal code that prohibits more than one "prayer meeting" a month in a private home, issued an order for the Reiters to cease and desist.

Reiter and her husband David, a clergyman, were shocked. The Bible meetings were no more than a private affair organized by Diane and had nothing to do with her husband's ministry work.

The couple had received some complaints about the ten cars or so parked in the street during the meetings, but this

number complied with city regulations and hardly constituted a neighborhood nuisance.

When the Reiters appealed the case to the zoning board
of appeals, its director told them that the problem wasn't
the cars—it was the fact that they were holding a prayer
meeting. If, on the other hand, the couple had been holding a book club meeting, then it probably would have been
"no problem."

The Reiters complained that other home gatherings,
such as "Monday Night Football" parties and home poker
games, were lawful, but not prayer meetings. They filed a
lawsuit in federal court charging the City of Denver with
violating their constitutional rights to religious freedom.

Eventually, the city settled out of court, paid $30,000
for all court costs, and agreed to change the code to allow
the prayer meetings to continue.

Maybe that city official should have sought spiritual
guidance before he took on the First Amendment and
squandered taxpayer dollars.

HOMEOWNERS WITH A VIEW, BEWARE
(Skamania County, Washington)

Brian and Joda Bea's home is perched on a gusty bluff
overlooking the picturesque Columbia River Gorge and
Oregon's 620-foot Multnomah Falls. There was, however,
a big problem with this house; it can be seen by too many
tourists.

Brian had inherited the land that had been homestead-
ed by his great-great-grandfather in the 1850s. The couple
spent several years winding through the permitting process and finally received a building permit from Skamania
County in 1997.

Two years later, when the home was nearly complete,

the Columbia River Gorge Commission, a bi-state government agency charged with regulating land use in the Gorge area, demanded that the Beas stop work. It ordered the couple to either tear down the house or move it to a less visible spot on their property and said that the home must be "visually subordinate to its landscape setting." In other words, blend in with the hillside.

Even though the home was an earth-tone color and more than a hundred trees had been planted to screen it from the river and highways below, the agency insisted that the home could be seen from several vantage points. One opposing attorney in the case complained that the house "sticks out like a sore thumb" but couldn't even find it in professional photographs taken from key viewing areas. In one instance, he even pointed to the wrong home.

Nevertheless, the commission stood fast. It justified its position by arguing that there is no limit on the time the agency can go back and revoke permits. The Beas had already spent more than $200,000 in constructing their home.

During their ordeal, the couple was unable to live in the home yet they continued to pay the mortgage as well as the cost of rent to live elsewhere. Brian was forced to work seventy hours a week just to make ends meet. At one point, the couple had to receive "permission" from the commission to live in their maintenance building next to the house to avoid further financial hardship. Meanwhile, the house had fallen into disrepair and had been vandalized.

The Beas persevered. After numerous administrative hearings and court battles that culminated in 2001, the Washington Supreme Court ruled in their favor. The case was settled and the Beas received a check from the commission for $305,000—and were finally able to move into their dream home.

Brian said, "It's nice to know that our new baby won't be homeless."

NOT IN THEIR BACKYARD
(Corona del Mar, California)

For the past thirty years, George and Sharlee NcNamees have enjoyed the magnificent view from their beach home overlooking Corona del Mar State Beach—until the California Coastal Commission came calling.

In 2001 the elderly couple was informed that two concrete tables, a storage locker, an outdoor shower, and a barbeque located in their backyard did not comply with the California Coastal Act. The commission contended that they needed a permit and said the "developments" cause a "perception of privatization" which discouraged the public's use of the beach.

The McNamees argued that they did not need a permit because the structures had been in place prior to the act being signed in 1975. The previous owner and neighbors provided sworn affidavits backing the McNamees' claim. The commission contended that it had aerial photographs showing that the improvements were built between 1986 and 1993. A photographic expert says the evidence is unclear.

Regular beachgoers have come to the couple's defense and said the improvements do not discourage their use of the beach. In fact, many of them testified that they have used the benches during their walks down the steep hill to the beach.

The commission also ordered the couple to restore the natural vegetation on a part of the bluff where the couple creates a flower garden that resembles an American flag each spring.

If the commission had their way, the McNamees would not be able to use their backyard. "It's amazing," says George. "You can pay your property taxes, but you can't use your backyard."

The McNamees have sued the commission and are

pursuing their case through the courts. Sharlee McNamee was quoting as saying, "Even if it weren't our property, we would do whatever we could to fight what is happening because every citizen has an obligation . . . to encourage the protection of individual private property rights."

Their neighbors and beachgoers will be pulling for them.

PAY ME NOW, OR PAY ME LATER
(Lee County, Florida)

Richard and Ann Reahard had the right to build homes and a marina on waterfront property with direct access to the Gulf of Mexico. The county, however, had a different use in mind for the Reahards' land and was willing to put the couple through a judicial meat grinder for fourteen years to defend its plan.

The Reahards owned forty acres that had been approved for 126 homes and a two-and-one-half-acre marina. A developer offered the couple $1.2 million for the parcel, but the deal collapsed in December 1984 when the county passed a comprehensive land plan—a document intending to manage growth more efficiently.

The parcel was "down-zoned" to a "resource protection area" and permitted only one single-family residence. The consequence: the value of the property plummeted to $40,000.

The Reahards then spent five years attempting to get the decision overturned. At one point the planning commission agreed to rescind the previous action, but the county commissioners denied it. The couple then offered to sell the property to the county for $600,000—half of what they were offered by the developer. The commission did not even respond to the offer.

The Reahards sued in state court. The judge ruled in favor of the Reahards and a jury awarded them $700,000. In what Richard described as "administrative mumbo-jumbo," the county was instrumental in frustrating the couple for more than ten years through a series of appeals and going back and forth between state and federal court.

Finally, by 1998, the county had run out of legal maneuvers and was ordered to pay the Reahards for the property and reimburse their legal expenses. The total taxpayer tab was more than $2.2 million—for land the county could have bought fourteen years earlier for $600,000.

KIDS AT PLAY, KEEP IT THAT WAY
(Clinton, Mississippi)

If necessary, cities must go to court to uphold ordinances to discourage unsightly development that could jeopardize the public's health, safety, and welfare—but for a tree house?

In early 1997, the Welch family sought and received permission from the City of Clinton, Mississippi, to build a two-story, Victorian-style tree house in the family's front yard. After receiving two anonymous complaints in 2002, Mayor Rosemary Aultman ordered the Welches to tear down the house by citing a provision of the city code that prohibits accessory structures in front yards.

Backed by 120 neighbors and supporters, the Welches appealed the decision to the city's planning and zoning board. The board, not willing to take on the vocal group, made no recommendation and simply passed the decision on to the town aldermen.

The aldermen—despite admitting that the tree house was not an eyesore and was well constructed—voted 6 to 1 to have it torn down. The Welches then requested a special exemption for the tree house. A poll conducted by a research group showed that 76 percent of registered voters in Clinton preferred the city to grant the exemption. Nevertheless, the application was denied—but the fight was not over.

The Welches appealed the decision in circuit court. The judge ruled in their favor and said that the city ordinances do not define a tree house. The city board then voted to appeal the decision before the Mississippi Supreme Court. In August 2003, the court sided with the Welches. It said that the city's ordinance was "unconstitutionally vague" and ordered that the tree house stay.

The Welches spent $28,000 in legal fees. The city spent around $30,000.

The civics lesson learned by the neighborhood children: priceless.

Part III

WETLANDS

13

REGULATING PUDDLES AND PONDS

"Ecologically speaking, the term 'wetland' has no meaning. . . . For regulatory purposes, a wetland is whatever we decide it is."

—ROBERT J. PIERRE, A FORMER FEDERAL GOVERNMENT OFFICIAL WHO HELPED WRITE THE 1989 U.S. WETLANDS MANUAL

Water is our most precious natural resource. No one would argue that polluters of our rivers, lakes, and oceans should be punished to the fullest extent of the law. However, overzealous government bureaucrats have gone too far by abusing their powers to regulate wetlands and intimidating landowners with fines and jail sentences.

The U.S. Environmental Protection Agency (EPA), one of the agencies authorized under the act to regulate wetlands, defines wetlands as "areas where water covers soil, or is present either at or near the surface of the soil all year or for varying periods of time during the year, including during the growing season."

Unfortunately, other agencies such as the U.S. Army Corps of Engineers, the U.S. Fish and Wildlife Service, the National Oceanic and Atmospheric Administration, and a host of state and local government regulators define wetlands differently. One thing that is not disputed is that these waters provide many benefits.

WETLANDS UNDER SIEGE

Wetlands—often referred to as giant kidneys—filter out pollutants, help replenish drinking water supplies, protect against floods by capturing and storing excess surface water, and provide habitat for more than a third of the country's threatened and endangered animal and plant species. In addition, they provide a number of agricultural products (such as timber and berry crops) and recreational opportunities for millions of Americans. However, wetlands across America have been declining at an alarming rate.

To address this crisis, former President George H.W. Bush's administration established a national goal of "no net loss" of wetlands and the Clinton administration set a goal to increase wetlands by 100,000 acres per year beginning in 2005. Even though there is evidence that destruction has slowed, it is estimated that 70,000 to 90,000 acres are still destroyed every year.

The largest and most recognized wetlands in the United States are Florida's Everglades National Park and the Louisiana Bayou Region. The Everglades encompass more than 4,000 square miles and the Louisiana bayous comprise a swath of marshes, islands, and swamps of more than 6,000 square miles. Both are under strain from development and are undergoing important restoration programs.

The Everglades' $15.6 billion, twenty-year plan has been hailed as the most ambitious environmental project in history. This "sea of grass" has been devastated over the past fifty years by attempts to redirect its flow to farms and cities. As many as sixty-eight animal and plant species are also considered to be at risk simply because there is a lack of fresh water.

The Louisiana delta, with one-third of the nation's run-off from the Mississippi River, has lost 1,900 square miles of wetlands since the 1930s. In the 1960s, the U.S. Army Corps of Engineers dredged fourteen major ship channels, while oil companies cut canals for pipelines and oil wells. Even though the state has launched an ambitious $14 billion program to recover some of the lost wetlands, the destruction continues at an estimated rate of twenty-five square miles per year.

THE POWER OF THE CLEAN WATER ACT

Wetlands do not, however, have to be as enormous as the Everglades or the Louisiana bayou to fall under the government's regulatory control. Waters considered to be large puddles or ponds also fall under this authority and have created bitter controversy between developers and environmental groups.

The National Association of Home Builders (NAHB) insists that the Army Corps of Engineers, the agency responsible for authorizing construction projects in wetland areas, is overburdening developers. Environmental groups, such as the Audubon Society, claim that the Corps of Engineers is instituting adequate protections. This tug of war has led to several lawsuits. At the heart of the debate is defining the Corps' regulatory jurisdiction and guidelines under the Clean Water Act.

In 1972 Congress amended the Water Pollution Act. It created a regulatory plan under Section 404 of the Clean Water Act to control the discharge, or release, of soils into wetland areas. These discharges result from projects such as channel construction or maintenance, building ports, using soils to create dry land (known as fill) near bodies of water, and for water control projects such as dams and levees.

Under these new rules, the Corps of Engineers was given jurisdiction over "waters of the United States" including wetlands "adjacent to navigable waters and their tributaries." The Corps' regulations then go on to define adjacent as "bordering, contiguous, or neighboring." This ambiguous definition has led to several lawsuits and some notable Supreme Court rulings.

The 1985 case of *U.S. v. Riverside Bayview Homes* involved a homebuilder that owned eighty acres of low-lying marsh near the shores of Lake St. Clair in Macomb County, Michigan. In 1976 Riverside began to place fill materials on its property in preparation for the construction of a housing development. The Corps of Engineers sought to halt the fill operation, citing jurisdiction, and filed suit in the district court.

The court sided with the Corps and ordered the activity to stop. Riverside appealed the ruling. The Court of Appeals reversed the lower court decision and said that the definition of the "waters of the United States" was too broad and that the regulations must be narrowly construed to avoid a taking under the Fifth Amendment of the Constitution. The Corps appealed to the U.S. Supreme Court.

The high court reversed the District Court of Appeals decision and said that the Corps had the authority to regulate the land since the wetlands were adjacent to navigable waters but did not define the extent to which the Corps could impose conditions on the development of the site.

In 1994 the Supreme Court heard a case, *Dolan v. the City of Tigard, Oregon,* where a city attempted to exert its authority over wetlands. The owner of a plumbing and electrical supply store in downtown Tigard wanted to expand her store and pave a gravel parking lot. As a condition of approval, the City Commission required her to dedicate property for flood control purposes, including a pathway for pedestrians and bicycles.

The Supreme Court ruled that the city's dedication requirement was an uncompensated taking of property under the Constitution. It further established that the city did not show a "reasonable relationship" between the construction of the improvements and legitimate regulatory goals. In addition, regulators must develop a balanced plan that does not overburden property owners.

In theory, the courts could then limit the government from forcing property owners to fund public improvements. In reality, however, what constitutes "reasonable" and "balanced" is highly subjective.

TWISTING THE COMMERCE CLAUSE

One of the most cited cases regarding wetlands has been *Solid Waste Agency of Northern Cook County (SWANCC) v. United States Army Corps of Engineers*. It is a good example of the lengths to which regulators will go to justify their authority and how activist judges attempt to legislate from the bench.

The case was argued before the U.S. Supreme Court in 2001 and involved a group of twenty-three suburban Chicago cities that wanted to develop a disposal site for nonhazardous material. The consortium, SWANCC, located an abandoned sand and gravel pit that had several permanent and seasonal ponds scattered throughout the property. The ponds ranged in size from less than an acre to several acres.

SWANCC contacted the Army Corps of Engineers to see if it needed a Section 404 permit since it had to fill some of the ponds. The Corps responded by saying that the ponds

were not considered wetlands and that it did not have jurisdiction. The Illinois Nature Preserves Commission, a
division of the state's Department of Natural Resources,
disagreed, citing that the commission had observed 121
types of migratory birds on the site.

The Corps then reversed its position and said it did indeed have federal jurisdiction. It justified its position by
saying that since the waters were of "natural character"
and migratory birds crossed state lines, the ponds were
then "waters of the United States." This legal maneuver is
now known as the "the migratory bird rule."

SWANCC attempted to work with the Corps and commission to minimize the disturbance of the birds, but it
could not satisfy the agency's concerns. SWANCC then filed
suit in District Court and lost. On appeal, the court took
up both the jurisdiction and the migratory bird rule questions.

The appeals court ruled that the Corps had jurisdiction
and—to the surprise of many observers—found that the
migratory bird rule was a "reasonable interpretation" of
the Clean Water Act because of the impact of the "destruction of the natural habitat of migratory birds" on interstate
commerce. It said that since millions of Americans cross
state lines and spend more than a billion dollars to hunt
and observe migratory birds, the ruling was consistent
with their interpretation of the commerce clause of the
Constitution that gives the federal government the power
to regulate interstate commerce. The Supreme Court disagreed.

The high court ruled that the Corps had overstepped
its legal authority under the Clean Water Act since the waters were not "navigable," and that the migratory bird rule
could not be cited to justify its position. This was a victory
for property rights advocates. However, there are several
environmental rulings—especially those involving endan-

gered species—where the commerce clause is twisted by appellate courts to defend extremist environmental agendas.

To avoid uncertain and protracted court battles (and since 75 percent of the remaining wetlands lie on private lands), some progressive property owners and conservationists are working in good faith with the Corps on the creation of innovative wetlands protection programs. These programs seek to achieve a balance between protecting private property rights and safeguarding the environment by encouraging, rather than discouraging, landowners to maintain wetlands. Unfortunately, the Corps, in many instances, continues to overstep its authority and is not being held accountable. In fact, very few of the ensuing controversies are ever reported.

One reason these disputes have not been highly publicized is because many small private property owners cannot afford to fight the regulators. Since the average cost to process a Section 404 Permit is $275,000 and can take more than two years to obtain, owners either abandon their plans to develop their property or agree to onerous conditions in order to move forward.

Developers of large projects are no exception. The cost to obtain approvals of large-scale projects can be much higher and take several years to secure. Since developers of these projects are deemed to have "deep pockets," the Corps, other regulators, and environmental groups are often able to extort large tracts of land and considerable sums of money in exchange for not creating a public controversy.

Developers consider this extortion a cost of doing business, but homebuilders (for example) pass these costs on to the consumer. This is a contributing factor in the lack of affordable housing in some areas of the country.

The courts have failed to provide a clear definition of

the Corps' authority. Bureaucrats and environmental litigants continue to use the Clean Water Act as a tool to control local land-use decisions. Many landowners throughout the country are required to go through the regulatory meat grinder to obtain permits—permits they shouldn't have to get in the first place.

14

A WETLANDS DESPERADO
(Navarre, Florida)

"If you sin, sin against God, not against the bureaucra-cy. God may forgive you. The bureaucracy never will."
—ANONYMOUS (AMERICAN)

Ocie Mills and his son were the first property owners in America to be imprisoned for polluting a wetland. What toxic material did these criminals dump that posed a danger to the public's health? Sand.

Ocie never shied away from hard work. As the oldest of seven children, he was forced to leave the eighth grade to help his father make a living in Holley, Florida. He worked as a carpenter, a fisherman, and a mate on an eighty-five-foot sailboat before serving in the military and marrying.

Ocie and his wife, Betsy, settled down in Minnesota to raise five children—two of their own and three adopted orphans. Ocie provided for his family by working as a service technician, leasing and operating a gas station, becoming a gasoline bulk supplier, and owning a restaurant—before moving back to the Florida panhandle in 1967 to go into the construction business with his brother.

The concrete plant and grading and paving businesses the brothers owned prospered and eventually employed as many as eighty-four people in the Fort Walton area. Life was good for Ocie and Betsy—until state environmental officers came calling one day in 1976.

NO ONE'S FOOL

On that fateful day, two men with the Florida Department of Environmental Protection (DEP) arrived at the Mills' beachfront home. Ocie had recently cleaned out a clogged drainage ditch on the property to get rid of a rattlesnake problem and the officers wanted to investigate whether or not the work violated state environmental laws.

The men flashed Ocie a green card and said it gave them the legal right to enter his property. Not only did Ocie have an inherent distrust of the government, but he also understood his constitutional rights. He, therefore, told the men that they could not come on the property unless they had a search warrant.

The men argued. At one point, an officer called Ocie a "redneck." But what really set Ocie off was when one of them said he had lost his property rights because he was violating state law.

There was a brief scuffle. Ocie, who was forty-one years old at the time, managed to wrestle both men (in their midtwenties or early thirties) to the ground. He grabbed a pistol from his truck and put it in his belt while Betsy called the sheriff.

The officers were arrested. The DEP then filed a suit against Ocie charging him with reckless display of a firearm and battery. He countersued for false arrest.

Confident he would win in court, Ocie decided to defend himself. The judge ruled in Ocie's favor and the state paid him $9,000 to settle. As a result, the DEP changed its policy and required officers to obtain a search warrant before entering private property. The green card hoax was stopped.

Ocie became a sort of private property rights hero in the Florida panhandle. His "David versus Goliath" case was highly publicized, and owners having trouble with the DEP or the U.S. Army Corp of Engineers sought his advice.

As word spread, he spoke before property groups throughout the country about how government regulators are violating our constitutional rights. He and his supporters formed the American Environmental Foundation, and he even ran, unsuccessfully, for the Florida House of Representatives.

THE SECOND RUN-IN

Ocie continued to spar with the regulators from time and time, but a heart attack in 1986 forced his decision to retire. He bought two lots on East Bay in Navarre, a small Gulf Coast town twenty-five miles east of Pensacola, with the intent to build a home for his son, Casey.

Prior to starting construction, Ocie contacted the DEP before cleaning out a four-foot-wide drainage ditch on the property. The standing water had attracted frogs and poisonous snakes and was infested with mosquitoes.

Officers visited the property. They placed blue flags in an area near the shore that they considered to be a wetland and told Ocie he could build anywhere else on the lot.

With all the necessary local building permits in hand, Ocie cleaned the ditch and then began construction. After bringing in two or three loads of clean sand, however, he received a cease-and-desist order from the U.S. Army Corps of Engineers demanding that he stop all construction activities.

The Corps claimed that it had jurisdiction over the ditch. Ocie contacted an official with the DEP who said the Corps had made a mistake and to move forward with the construction. After he ignored another cease-and-desist order from the Corps, however, armed government officials, including the FBI, came to the site.

The men threatened to arrest workers who were dumping sand on the property and ordered them off the site. It was all a set-up, Ocie believed—payback time for all his harsh criticism of environmental agencies over the years.

A special investigator and two attorneys were brought in from the Corps' Atlanta office. The investigation went on for four months, and eventually Ocie was indicted by a grand jury for "polluting the navigable waters of the United States." Although his son, Casey, did not take part in this "illegal activity," he, too, was indicted because Ocie had placed his name on the property deed.

Ocie couldn't believe what was happening. He reread the U.S. Constitution, reexamined his building permits, and decided to fight it. Since he had successfully defended himself in the past in state court, he figured he would do the same in federal court.

Despite U.S. District Judge Winston Arnow's recommendation that he hire a lawyer—explaining that federal court rules and procedures are much stricter than in state court—Ocie and Casey decided to go it alone. After all, the law was clearly on their side.

They prepared for the case by amassing documents, lining up witnesses, and looked for problems in the prosecutor's case. Their strategy was to convince the jury that they were just little guys trying to swim through a sea of red tape. Ocie soon discovered, however, that federal court was not a friendly place for amateur lawyers.

IN OVER THEIR HEADS

During the trial, Ocie and his son were painted as dangerous criminals who ignored the government's cease-and-desist orders and knowingly polluted the nation's waters. Richard Windsor, an assistant general counsel for the DEP, even compared Ocie to Humpty-Dumpty who was known for creating his own version of reality.

When it was his turn to speak, Ocie presented the building permit issued by the county and notes of his phone conversations with the state DEP official who had told him to ignore the Corps' cease-and-desist orders. Arnow, an 82-year-old judge whom Ocie claims was hard of hearing and senile, ruled that none of his evidence was admissible because the county and state's authority was superseded by the federal government.

Unable to present his evidence, Ocie could see the writing was on the wall. Ocie, 64, and Casey, 41, were convicted. They were sentenced to twenty-one months in federal prison, six months' probation, fined $5,000 each, and ordered to remove nineteen loads of sand from the property.

Paul Craig Roberts, a *Washington Times* reporter who followed the case, would later obtain documents through the Freedom of Information Act that proved that the government was out to get Ocie. One letter from an Army Corps official to the DEP strongly urged prosecutors to bring criminal charges against Ocie because he had furnished advice to others "with the intent to subvert the Corps' regulatory program."

Ocie and his son began to serve their time. While in prison, they were interviewed by a local television station. In a segment titled "The Sandman in Jail!" Ocie said, "There is no hope to recuperate the damage that's took place. No. But that don't mean that I'm not gonna fight . . . because our freedom in our country depends on it."

Ocie and Casey filed an appeal with the Eleventh Circuit Court but lost. In 1990 they hired Ron Johnson, a Pensacola attorney, and filed what is known as a *Bivens action*—a type of claim used to hold federal employees personally liable for actions they take which deprive someone from exercising their constitutional rights.

In the suit, they requested that their sentences be dismissed. The District Court, however, denied their motion. The Appeals Court agreed with the ruling, and eventually the U.S. Supreme Court denied a request to hear the case.

By November 1991 Ocie and his son had completed their prison time and probation but decided to fight the part of their conviction that required that they lower the elevation of the property by 11 inches—doing so would render the property unusable because the majority of the lot would be under water at high tide.

In 1993, with a new federal judge, Roger Vinson, on the bench, Ocie and his son went back to court to determine the "original wetland elevation" and attempted to have their felony convictions taken off their records. The judge ruled that the property was "probably never a wetland for the purposes of the Clean Water Act." Even though he believed Ocie and Casey were innocent, the judge could not set aside the convictions since the issue of whether or not the land was a wetland was not presented at the original trial.

JURORS ACTING BADLY

A few years went by. Then one evening after a news program aired Ocie's story, Ocie received a call from Quentin Wise, one of the jurors in the original case. Wise had watched the program and said, "Mr. Mills, you were a guilty man before you even had a trial."

Wise informed Ocie that the jury foreman, Thomas J. Smith, had told jurors during the trial that his sons worked for the DEP, that he had learned that Ocie had threatened officers with a gun in the past, and portrayed him as a terrible polluter. Wise confessed that he and the other jurors had been intimidated by Smith and that they had agreed to convict Ocie and his son.

This was a significant turn of events. The Sixth Amendment of the Constitution states: "In all criminal prosecutions, the accused shall enjoy the right to a speedy and public trial, by an impartial jury." Since the jury had been "tainted," Ocie believed he had a good shot at overturning the sentences and restoring his reputation.

Wise agreed to provide a sworn statement that Thomas Smith had introduced evidence during jury deliberations that was not presented at trial. Ocie then phoned Smith.

Smith's wife answered the phone and told Ocie that her husband had died more than a year earlier. According to Ocie, she told him that "he went to his grave hating what he had done to you and your son."

Another juror, Clyde Smith, backed Wise's story. He said, "We were more or less led to believe [the sentence] wouldn't be any more than a slap on the wrist, and that nothing would come of it."

In April 1996 Ocie's attorney filed a motion called a *Coram Nobis*, a Latin term for "an error before us." This type of maneuver is often used as a last resort by criminal defendants who allege a fundamental error or misconduct in the case that influenced the conviction.

The government sought to dismiss the motion, but the district judge agreed that Ocie had a valid claim. The appeals court, however, disagreed and said that—even though there was juror misconduct—it did not rise to a high enough level to overturn the sentences.

In January 2001 Ocie appealed to the U.S. Supreme Court, but it declined to review the case, saying that "the misbehavior or partiality of jurors" is not a fundamental error in the case.

Ocie's fifteen-year $200,000 legal battle had come to an

end. Betsy, who had been in failing health ever since Ocie and his son were sentenced to prison, passed away in 2003. Two years later, Ocie's request to President Bush for a presidential pardon was denied.

The convicted felon now lives on Social Security and veterans' benefits. He divides his time between his daughter's home in Pensacola, Florida, and a log cabin in Minnesota where he runs a hunting camp and operates a small sawmill. His troubles with government bureaucrats, however, are not over.

In January 2007 Ocie received a Wetlands Conservation Act cease-and-desist order from the Minnesota Department of Natural Resources. The agency discovered that he had cleared a road leading to his sawmill. The order demanded that he "stop all work, conduct no further work, and take immediate corrective action to stabilize the site from imminent erosion or restore water flow." It also required that he apply for a permit.

Although there is no evidence to suggest conservation officers are out to get him for his past transgressions in Florida, the 73-year-old does not want to tempt fate. At last report, Ocie is trying to sell his sawmill.

15

THOSE DAM BUREAUCRATS!
(Pierson, Michigan)

"In a bureaucratic system, useless work drives out useful work."

—**MILTON FRIEDMAN, ECONOMIST**

In **1997 one** of Stephen L. Tvedten's neighbors reported to the Michigan Department of Environmental Quality (DEQ) that he had illegally built two creek dams. The agency then sent a cease-and-desist letter ordering him to remove the debris. Maybe the DEQ should have looked into the matter rather than simply taking the word of Tvedten's neighbor.

The following is the letter from an official at DEQ that was sent via certified mail to Tvedten's property address:

Mr. Ryan DeVries
2088 Dagget
Pierson, MI 49339
SUBJECT: DEQ File No. 97-59-0023—1 T11N, R10W, Sec. 20, Montcalm County

It has come to the attention of the Department of Environmental Quality that there has been recent unauthorized activity on the above referenced parcel of property. You have been certified as the legal landowner and/or contractor who did the following unauthorized activity: Construction and maintenance of two wood debris dams across the outlet stream of Spring Pond.

A permit must be issued prior to the start of this type of activity. A review of the Department's files show that no permits have been issued. Therefore, the Department has determined that this activity is in violation of Part 301, Inland Lakes and Streams, of the Natural Resource and Environmental Protection Act, Act 451 of the Public Acts of 1994, being sections 324.30101 to 324.30113 of the Michigan Compiled Laws annotated.

The Department has been informed that one or both of the dams partially failed during a recent rain event, causing debris and flooding at downstream locations. We find that dams of this nature are inherently hazardous and cannot be permitted. The Department, therefore, orders you to cease and desist all unauthorized activities at this location and to restore the stream to a free-flow condition by removing all wood and brush forming the dams from the stream channel. All restoration work shall be completed no later than January 31, 1998.

Please notify this office when the restoration has been completed so that a follow-up site inspection may be scheduled by our staff. Failure to comply with this request, or any further unauthorized activity on the site, may result in this case being referred for elevated enforcement action. We anticipate and would appreciate your full cooperation in this matter.

Please feel free to contact me at this office if you have any questions.

Sincerely,
David L. Price
District Representative Land and
Water Management Division

The following is Stephen L. Tvedten's reply:

Re: DEQ File No. 97-59-0023; T11N, R10W, Sec 20;
Montcalm County

Dear Mr. Price:
Your certified letter dated 12/17/97 has been handed to me to respond to. You sent out a great many carbon copies to a lot of people, but you neglected to include their addresses. You will, therefore, have to send them a copy of my response.

First of all, Mr. Ryan DeVries is not the legal landowner and/or contractor at 2088 Dagget, Pierson, Michigan; I am the legal owner and a couple of beavers are in the (State unauthorized) process of constructing and maintaining two wood "debris" dams across the outlet stream of my Spring Pond.

While I did not pay for or authorize their dam project, I think they would be highly offended that you call their skillful use of natural building materials "debris." I would like to challenge you to attempt to emulate their dam project any dam time and/or any dam place you choose. I believe I can safely state there is no dam way you could ever match their dam skills, their dam resourcefulness, their dam ingenuity, their dam persistence, their dam determination and/or their dam work ethic.

As to your dam request the beavers first must fill out a dam permit prior to the start of this type of dam activity, my first dam question to you is "Are you trying to discriminate against my Spring Pond Beavers or do you require all dam beavers throughout this State to conform to said dam request?" If you are not discriminating against these particular beavers, please send me completed copies of all those other applicable beaver dam permits. Perhaps we will see if there really is a dam violation of Part 301, Inland Lakes and Streams, of the Natural Resource and Environmental Protection Act, Act 451 of the Public Acts of 1994, being sections 324.30101 to 324.30113 of the Michigan Compiled Laws annotated.

My first concern is whether or not the dam beavers are entitled to dam legal representation? The Spring Pond Beavers are financially destitute and are unable to pay for said dam representation, so the State will have to provide them with a dam lawyer. The Department's dam concern that either one or both of the dams failed during a recent rain event causing dam flooding is proof we should leave the dam Spring Pond Beavers alone rather than harassing them and calling their dam names. If you want the dam stream "restored" to a dam free-flow condition, contact the dam beavers; but if you are going to arrest them (they obviously did not pay any dam attention to your dam letter—being unable to read English) be sure you read them their dam Miranda rights first.

As for me, I am not going to cause more dam flooding or dam debris jams by interfering with these dam builders. If you want to hurt these dam beavers; be aware I am sending a copy of your dam letter and this response to PETA [People for the Ethical Treatment Of Animals].

If your dam Department seriously finds all dams of this nature inherently hazardous and truly will not permit their existence in this dam State. I seriously hope you are not selectively enforcing this dam policy, or once again both I and the Spring Pond Beavers will scream prejudice!

In my humble opinion, the Spring Pond Beavers have a right to build their dam unauthorized dams as long as the sky is blue, the grass is green and water flows downstream. They have more dam right than I to live and enjoy Spring Pond.

So, as far as I and the beavers are concerned, this dam case can be referred for more dam elevated enforcement action now. Why wait until 1/31/98? The Spring Pond Beavers may be under the dam ice then, and there will be no dam way for you or your dam staff to contact/harass them then.

In conclusion, I would like to bring to your attention a real environmental quality (health) problem; bears are actually defecating in our woods. I definitely believe you should be persecuting the defecating bears and leave the dam beavers alone. If you are going to investigate the beaver dam, watch your step! (The bears are not careful where they dump!)

Being unable to comply with your dam request, and being unable to contact you on your dam answering machine, I am sending this response to your dam office.

Sincerely,
Stephen L. Tvedten
cc: PETA

The DEQ dropped its investigation after an employee inspected the property. A department spokesman said, "It probably would have been a good idea to do an inspection before we sent the notice."

A WETLANDS NIGHTMARE
(Midland, Michigan)

"If even one molecule, one drop, a trickle of water flows from point 'a' to a navigable water, it's subject to federal control . . . [even] from twenty miles away."

—REED HOPPER, PACIFIC LEGAL
FOUNDATION REGARDING
RAPANOS V. UNITED STATES

John Rapanos is a hardened criminal. For more than eighteen years, the Michigan grandfather has been the target of a crusade by the federal government to put him behind bars and fine him as much as $13 million. His crime? He moved sand around on his own property.

Rapanos is the 70-year-old son of a Greek immigrant fruit peddler. His father had gone broke in Chicago during the Depression and had settled in Midland, Michigan. John became a successful developer of subdivisions and commercial sites. Life was good—until 1988, when he decided to develop a 200-acre cornfield he had owned since the early 1950s.

According to court documents, Rapanos contacted the Michigan Department of Environmental Quality (DEQ) to inspect a portion of the property he suspected could be categorized as wetlands. He was informed that it was likely that a portion of the property, known later as the Salzberg site, could be a regulated wetland.

Rapanos then hired a wetlands consultant to evaluate the site. The consultant concluded that there were forty-eight to fifty-eight acres of wetlands on his property. Rapanos disagreed strongly with the findings.

Allegedly, Rapanos ordered the consultant to destroy the report as well as all references to it. The consultant later testified that when he refused, Rapanos threatened to "destroy him" if he didn't comply. In addition, the consultant stated in court that Rapanos claimed he would bulldoze the site himself—regardless of the study's conclusions.

100-YEAR-OLD DITCHES

In April 1989 Rapanos directed workers to clear the land and fill in low spots with dirt from the property. In August the DEQ made an attempt to inspect the site. Rapanos refused them access.

Three months later, officials from the DEQ returned. They concluded that ditches had been dug and that the site had been drained. Rapanos told the DEQ officers he had purchased the land from the State of Michigan. The drainage ditches on all sides and through the middle of the property had been in place when he bought it. In addition, his understanding was that the County Drain Commission had begun digging the ditches on the land as early as 1904!

The DEQ also alleged that Rapanos had hauled in 300,000 yards of dirt to fill the site. Considering that an average truckload contains 10 cubic yards, that would mean around 30,000 truckloads. Rapanos claimed that no fill had been brought to the site.

The DEQ pressed forward with its claims. Soon thereafter, the federal Environmental Protection Agency (EPA) issued Rapanos a cease-and-desist order to stop filling the Salzberg site and other parcels. But the feisty developer would not back down and kept grading his property. The matter was then referred to the U.S. Department of Justice and in 1993 a grand jury indicted Rapanos.

By July 1994, both criminal and civil charges were brought against Rapanos in District Court for illegally filling protected wetlands in violation of the Clean Water Act. Rapanos' lawyers argued that the government did not have jurisdiction over his alleged wetlands because portions of the land were more than twenty miles from the nearest navigable waters and their tributaries.

The attorneys cited the landmark 2001 Supreme Court ruling known as *Solid Waste Agency of Northern Cook County (SWANCC) v. United States Army Corps of Engineers*. The decision was an attempt to define the Corps' jurisdiction. However, later appellate court decisions seemed to disagree with the intent of *SWANCC*, thereby failing to establish a clear standard.

During the criminal trial, the testimony from Rapanos' consultant was damaging. Testimony made by Russ Harding, the former director of Michigan's DEQ, however, seemed to substantiate Rapanos' position.

A JUDGE WITH SENSE

Harding had visited the site and had dug forty to fifty holes, each four feet deep, at various spots around the property. He found no trace of water and no evidence of fill dirt, as the government had alleged. In fact, had Rapanos brought in as much dirt as the government claimed, the site would have been six feet higher than the surrounding farms. Other experts confirmed the findings.

Although evidence that Rapanos had not dug the ditches

in the first place was confirmed and there was no evidence that fill was brought to the site, the prosecutors pressed on.

Despite evidence to the contrary, a jury sided with the government. It concluded that fifty-four acres of the property had been filled and Rapanos had violated the Clean Water Act. In addition, they found that Rapanos had concealed evidence, but District Court Judge Lawrence Zatkoff ruled that the prosecutors had wrongly accused Rapanos of that and set it aside. The Justice Department appealed.

The appeals court reversed Zatkoff's ruling and sent the case back to District Court and demanded that Zatkoff sentence Rapanos. At the sentencing hearing—which followed the sentencing of an illegal immigrant for drug trafficking—the judge signaled his disgust at the Justice Department's prosecution of Rapanos and said: "So here we have a person who comes to the United States and commits crimes of selling dope and the government asks me to put him in prison for ten months. And then we have an American citizen who buys land, pays for it with his own money, and he moves some sand from one end to the other and the government wants me to give him sixty-three months in prison. Now if that isn't our system gone crazy, I don't know what is. And I am not going to do it."

Instead, Zatkoff sentenced Rapanos to 200 hours of community service, three years' probation, and a fine of $185,000. The government appealed and the appeals court, at the urging of the Justice Department, ordered the judge to imprison Rapanos for ten months. Rapanos appealed to the U.S. Supreme Court.

The high court sent the case back to the district court to reconsider the case. The district court then found that the government lacked jurisdiction in the case and set aside the conviction. But the government would not give up.

Prosecutors appealed the ruling. The appeals court reversed the district court's decision, reinstated the conviction, and sent it back to Judge Zatkoff for sentencing.

SOMETHING IS OUT OF WHACK

On March 15, 2005, Zatkoff ruled on the sentencing issue. His written opinion stated that in the State of Michigan, defendants in environmental cases had rarely been sentenced to prison terms. He cited a case in which two Macomb County developers clear-cut, plowed, and destroyed eight acres of forested wetlands between 1995 and 1999—despite repeated warnings. They were sentenced to one year probation, fined $5,000, and ordered to fully restore the wetlands.

Another case he cited involved the Department of Justice's prosecution of BP, a major oil company, for discharging pollutants from a facility directly into the Delaware River over a six-year period. These violations were done knowingly, but the Justice Department only prosecuted BP in a civil case, not both civilly and criminally as in the Rapanos case. Ultimately, BP was fined and merely told to clean up its facility.

Judge Zatkoff also made reference to what is arguably the most notorious environmental disaster of the twentieth century—the *Exxon Valdez* oil spill in 1989. The tanker struck a reef in Alaska and dumped 10.8 million gallons of crude oil in the water and created a huge oil slick that killed tens of thousands of animals. The slick impacted 1,300 miles of coastline and reportedly killed approximately 250,000 birds, 2,800 sea otters, 300 seals, and 250 bald eagles, while crippling the local fishing industry.

It was determined that the captain of the oil tanker was drunk the night of the spill and that his negligence was the cause of the accident. He was sentenced to serve 1,000 hours of community service over five years—no prison time or fine.

Zatkoff, therefore, concluded that imprisonment was not justified and imposed the original sentence of three years' probation and a fine of $185,000, which Rapanos had already satisfied.

The judge condemned the prosecutors for making the case personal and, according to the Associated Press, accused the government of going overboard in the case because Rapanos is a "disagreeable" person. "This is the kind of person," said the judge, "the Constitution was passed to protect."

"John Rapanos has been vindicated," said Pacific Legal Foundation principal attorney Reed Hooper, who represented Rapanos. "Judge Zatkoff has confirmed what we have been saying all along—this case has always been about government power, not protecting wetlands."

BEING MADE A TARGET

In addition to the numerous criminal proceedings in the case, the civil case was making its way through the court at the same time. The district court (not Zatkoff's court) also agreed that Rapanos filled fifty-four acres. Government prosecutors sought: 1) a $10 million fine; 2) a $3 million in mitigation fee to buy eighty-one acres of wetlands to replace what Rapanos allegedly destroyed; and 3) up to fifty-six months in prison.

By this time Rapanos had endured 17 years of litigation. He had spent $1.5 million in legal fees and had allegedly lost $4 million in revenue since the Department of Justice had tied up his property. All this, of course, begs the question: Why did the government want to punish Rapanos?

Russ Harding, the former Michigan regulator now with the Mackinac Center, a free market think tank, claims that the Justice Department and EPA wanted to target a successful developer and create a high-profile case. In other words, to make an example of him.

Rapanos appealed the civil case. The appeals court ruled for the government. Rapanos vowed to fight on and decided to appeal to the U.S. Supreme Court.

BATTLE LINES DRAWN

On February 21, 2005, the Pacific Legal Foundation filed its petition. The main question they posed: Does the Clean Water Act extend to nonnavigable wetlands that do not even lie alongside a navigable waterway?

The Act specifies that the U.S. Army Corps of Engineers is given jurisdiction over "waters of the United States" including wetlands "adjacent to navigable waters and their tributaries." The Corps' regulations go on to define adjacent as "bordering, contiguous, or neighboring." In the Rapanos case, however, the navigable waters were twenty miles away.

The government responded and cited an appeals court decision known as *United States v. Denton*, which said that the Army Corps of Engineers has jurisdiction over nonnavigable tributaries if a connection or "nexus between a navigable waterway and its nonnavigable tributaries" is made.

On October 11, 2005, the Supreme Court agreed to hear the case. In addition, it was announced that another Michigan wetlands case, *Carabell v. U.S. Army Corps of Engineers*, would be heard at the same time.

In *Carabell*, a developer wanted to build a 112-unit condominium project on nineteen acres by filling in fifteen acres of forested wetlands. Unlike the property in *Rapanos*, these wetlands were adjacent to a lake. The Army Corps denied a permit, and Carabell brought suit alleging that wetlands are not covered under the Clean Water Act.

Not unexpectedly, Michigan's Attorney General Mike Cox supported the government's position in both cases. He filed a "friend of the court" brief, and in an interview said, "Michigan is a Great Lakes state. But those lakes will only stay great if we protect the rivers, streams, and wetlands that flow into them."

A coalition of environmental and public health groups also filed briefs. They argued for continued protection of streams and wetlands from industrial pollution and wrote that the cases concerned not only the filling of wetlands but also the regulation of "discharges of sewage, sediment and toxic chemicals such as cyanide from factories."

The National Wildlife Federation (NWF), joining with several conservation and sporting groups such as Ducks Unlimited and the American Sport Fishing Association, also filed a brief. An NWF press release stated, "The lower court properly recognized that the Clean Water Act was intended by Congress to broadly protect America's waters. If the Supreme Court reverses the lower court findings, it would leave our children a sad legacy of lifeless and polluted wetlands, streams, lakes and rivers."

Not unexpectedly, realtor organizations, homebuilders, the American Farm Bureau, cattle ranchers, the Chamber of Commerce of the United States, and the American Petroleum Institute wrote briefs in support of Rapanos and Carabell. In addition, hundreds of water agencies weighed in, including the nation's largest, the Metropolitan Water District of Southern California.

"These agencies," said Reed Hooper, are "on the front lines of providing clean water for tens of millions of Americans" and "have seen first-hand the abuse of the law by the federal government." He added that, in "the federal government's view, every drop of water in the country is within its reach," including "drainage ditches, concrete water run-offs, even pipes." During oral argument on February 21, 2006, Justice Antonin Scalia seemed to agree.

A CONFUSING DECISION

Scalia, considered to be one of the more conservative justices, was questioning Paul Clement, the government's Solicitor General, about whether storm drains should be under the jurisdiction of the federal government. Clement replied, "The Corps has not drawn a distinction between man-made channels or ditches and natural channels or ditches. And, of course, it would be very absurd for the Corps to do that since the Erie Canal is a ditch."

Scalia responded: "I suggest it's very absurd to call that waters of the United States. It's a drainage ditch," and he compared it to "a gutter in the street."

On June 19, 2006, the Supreme Court issued its ruling. The vote was 5 to 4 in favor of Rapanos and Carabell. Chief Justice Roberts, Justices Scalia, Alito, Thomas, and Kennedy were in the majority while Justices Stevens, Breyer, Ginsburg, and Souter, considered to be the liberals on the court, dissented. However, Justice Kennedy issued his own opinion, which in effect becomes the "controlling opinion."

In it, Kennedy established a new test stating that the Army Corps can regulate only wetlands that have a "significant nexus" to a major waterway. The opinion drew sharp criticism from both sides.

Chief Justice Roberts said that the result was confusing and that "lower courts and regulated entities will now have to feel their way on a case-by-case basis." Justice Stevens, writing the minority opinion, said that the opinion will create more uncertainty and additional work for regulators and landowners and "will probably not do too much to diminish the number of wetlands covered by the act in the long run."

Reed Hooper of the Pacific Legal Foundation seemed to be cautiously optimistic: "The court is clearly troubled by the federal government's view that it can regulate every pond, puddle, and ditch in our country." He added that the decision "represents a good step toward common sense regulation."

However, Joan Mulhern with Earthjustice, an environmental group based in Oakland, California, said, "I'd hesitate to say I agree with it or say I don't agree with it because I'm not sure what it means."

The court ordered the Rapanos case sent back to lower courts to be reconsidered under the new "rules." Although the decision eliminated Rapanos' liability of $13 million in fines, as well as prison time, the Justice Department prosecutors will get another shot at him.

The legal ordeal of John Rapanos is far from over—nor is the controversy over the government's right to regulate every trickle of water.

17

STORY SNAPSHOTS

DUMPED ON OVER A DUMP
(Morrisville, Pennsylvania)

The day after Thanksgiving in 1990, John Pozsgai, an immigrant from Hungary, began serving a three-year federal prison sentence for violating the Clean Water Act. His crime: cleaning up a 30-year-old dump.

Pozsgai purchased a property across from his home—full of discarded cars, scrap metal, and thousands of old tires—in order to expand his truck repair business. Although he was told by Pennsylvania's Department of Environmental Resources (PDER) prior to buying the property that it was not part of the National Wetlands Inventory, the U.S. Army Corps of Engineers classified the land as a wetlands and told him to stop cleaning up the site. The agency cited a stormwater ditch on the property that had to be maintained.

Pozsgai told the agency that the ditch had been dug in 1936 and was required to be maintained by the local Township of Morrisville. The PDER responded by suing. It also referred the case to the Environmental Protection Agency (EPA), which then referred it to the Department of Justice for criminal prosecution.

Pozsgai was eventually arrested by two EPA officers. They searched his house for weapons and found only kitchen knives.

Even though the Township of Morrisville eventually recognized its responsibility to maintain the drainage ditch, a federal prosecutor in the case said the prison sentence and $200,000 fine sent "a message to the private landowners, corporations, and developers of this country about President [George H.W.] Bush's wetlands policy."

One EPA official, however, tried to portray the agency as being "sensitive to the interests and concerns of landowners" and that its wetlands permitting "involves more than strict technical considerations and must include the sensitivity to the rights and expectations of all our citizens."

Pozsgai's fine was reduced to $5,000, but he and his family could not pay the property taxes and were forced to declare bankruptcy.

The controversy surrounding Pozsgai continued. After he had served his sentence, three environmental groups went to court to ask for the right to sue Pozsgai and force him to "restore" the site. The *Philadelphia Inquirer* wrote that these groups believe "the Army Corps of Engineers and the U.S. Attorney's office have not been tough enough."

SQUASHING A PROPERTY OWNER'S RIGHTS (Old Orchard Beach, Maine)

It looked like the land Gaston Roberge and his wife bought in 1964 to provide for their retirement was going to pay off—until the Army Corps of Engineers said that their lot was a wetland.

In 1986 the Roberges were offered $440,000 by a developer for their 2.8-acre lot that was just 300 feet from the Atlantic Ocean. The buyer, however, backed out of the deal when the Army Corps declared that the couple was in violation of the Clean Water Act since the lot had been illegally filled with dirt.

Apparently the violation stemmed from when the Roberges had accepted 200 loads of excess dirt from town fifteen years earlier. Even though all the proper permits had been obtained at the time, the Corps still claimed a violation. Coincidentally, this *illegal* activity took place *prior* to the passage of the Clean Water Act.

The Roberges then spent $50,000 on consultants to dig holes and write reports. They applied for an "after the fact" permit which would have allowed the property to be developed. The Corps, however, had no intention of approving it.

In fact, among several internal memos that were later discovered, one disgruntled Corps official said, "He [Roberge] has been given the written and oral 'dark cloud,' but he insists that we go on with his application." After three years, the permit was denied.

With the support of a Maryland-based property rights group, The Fairness to Landowners Committee, the Roberges sued the Army Corps and alleged that their property had been taken. Eventually, it was discovered that the Corps never adequately surveyed the property. In addition, internal Corps documents were uncovered indicating that the agency was using the case as a way to slow development in the area. A Corps official, Jay Clement, wrote: "Roberge would be a good one to squash and set an example—Old Orchard is heating up these days." He signed the memo "Formerly the Maytag Repairman," an apparently humorous reference to the fact there was little enforcement activity going on at the time.

Clearly embarrassed by the remarks, the Corps opted to settle the eight-year battle. They conceded that the land was never a wetland and paid the elderly couple $338,000—the first time the federal government has ever paid for a temporary wetland taking.

The only action taken against the Corps' Clements and

other officials was requiring them to attend a sensitivity training course. A spokesman for the Corps warned against judging the agency on the basis of a single memorandum written in a moment of "levity." He added, "We [at the Corps] wear white hats."

A PERMIT . . . BUT NOT REALLY
(Coon Rapids, Minnesota)

Helen and William Cooley owned a thirty-three-acre lot with plans to subdivide it and sell home sites. They went through a lengthy local approval process and then in 1989 asked the Army Corps if there were wetlands. The answer was yes—the entire parcel.

The Cooleys challenged the Corps' findings but eventually applied for a permit to fill twenty-six acres of the property. Four years later, the Corps denied the permit. The agency's letter said that the proposed development would produce "an unacceptable degradation of a valuable wetland resource" and that "the issuance of the fill permit would be contrary to the public interest."

The Cooleys sued the agency and alleged that their property had been taken without just compensation. Concerned it could lose the case, the Corps then sent the Cooleys a letter suggesting that they might be able to obtain a permit to fill only fourteen acres. The Cooleys declined the offer. As a legal maneuver to limit (or eliminate) liability, the Corps issued the permit anyway. Again, the Cooleys rejected it and maintained that the Corps could not force them to accept a permit for which they had never applied.

In 1996, just ten days before going to trial, the Corps issued a "provisional permit" for the entire property and included wording that it was "not valid" and stated "do not begin work." In a bold move, the Corps' attorneys argued at trial that the issuance of the permit nullified the Cooleys' takings claim.

The U.S. Court of Federal Claims did not buy the argument. After a thirteen-year legal struggle, the Cooleys were awarded more than $2 million in compensation.

NO GOOD DEED GOES UNPUNISHED
(Sonoma Valley, California)

Sam and Vicki Sebastiani thought they were doing a good thing when they decided to turn part of their 175-acre winery into a wetlands for waterfowl. Instead, government agencies punished them.

In addition to being a member of one of California's most noted wine families, Sam Sebastiani was also a dedicated conservationist. He decided to convert more than half of his property that flooded each rainy season into wetlands. To do so, he enlisted the help of California Ducks Unlimited and biologists at the California Department of Fish and Game.

Since 1993, those wetlands have become a significant wintering site. More than 156 species of birds have been recorded on the site—and on a single day more than 10,000 waterfowl have been counted. The project has been a great success—despite the intervention of eight local, state, and federal agencies nearly regulating it to death.

The undertaking should have cost only $50,000 and taken sixty days to complete, but due to overlapping regulations and onerous conditions the price tag was inflated to $181,000 and eight months to completion. The U.S. Army Corps of Engineers was particularly difficult.

Rather than viewing it as a wetlands restoration project, the Corps considered it wetlands destruction. It claimed that the construction of a 1,500-foot levee, necessary to create the wetlands as well as to protect a neighbor's farm from flooding, destroyed existing wetlands. The agency then required the Sebastianis to build another four-acre wetlands on another section of the property.

An official with the California Department of Fish and Game said the "exuberance" of federal regulators in enforcing the letter of the law "is going to be counter-productive and discourage" property owners from undertaking such projects. Sam Sebastiani would clearly agree.

SUNK BY A MUD PUDDLE
(Long Island, New York)

The Town of Southampton, New York, found a way to stop Walter Olsen and his wife from building on a lot. It told the couple that they had to set aside a chunk of their land to protect a mud puddle—a puddle that was wet only one to two weeks out of the year.

In order to plan for their eventual retirement, the Olsens purchased a 1.5-acre waterfront lot in 1987 with plans to build and lease out a small restaurant. Prior to buying the land, the couple was aware that a portion of the land was considered a saltwater marsh and would have to be protected. Therefore, as a condition of approval to build the restaurant, they agreed to set aside one-third of the property from development.

They redrew their plans and continued with the application process for another three years. When the Olsens were on the verge of gaining final approval, however, the town designated a portion of their property as freshwater wetlands.

According to the couple, there was absolutely no evidence presented by the town, a state environmental agency, or anyone else to prove that a freshwater wetlands existed.

The town, however, declared that a man-made puddle caused by runoff from a state highway constituted a wetlands and said another third of the property must be protected.

Plans were redrawn to accommodate the puddle, but there was not enough land remaining to build the restaurant. The Olsens sued the City of Southhampton but lost in court. The city had successfully argued that there were twenty-seven other uses to which the property could be put. Walter Olsen said, "You can imagine how old I'd be if I went through the twenty-seven other uses if this application took ten years."

After seven more years of unsuccessful attempts to develop the property, the couple had no choice but to sell the lot to the town for $150,000 to pay their legal and consulting fees—although the property's assessed value was $350,000.

BOGGED DOWN BY THE EPA
(Carver, Massachusetts)

For the past twenty years—as part of a federal wetlands crackdown—the U.S. Environmental Protection Agency (EPA) has been targeting cranberry farmers. Many farmers, without the resources to fight the agency, have been forced to comply with onerous regulations, but not the Johnson family.

Charles Johnson and his family have farmed cranberries since the 1920s. At various times between 1979 and 1999, they constructed and expanded bogs that have always been used for that purpose. The EPA, however, claimed that the Johnsons violated the Clean Water Act by not obtaining a federal permit before converting these "wetlands."

Based on a landmark Supreme Court decision, federal

regulatory power is limited to "navigable waters," such as a river or lake that can be used for shipping or other commerce, and wetlands immediately adjacent to such waters. Although none of the Johnsons' sites are located within twenty miles of a navigable waterway, the EPA contends water on their property *could* connect to a river by means of nonnavigable rivers, ditches, and streams.

Charles Johnson says the "stream" that regulators contend links the property to a river is only four feet wide and often dries up. "It's what they describe as navigable, but to them it's navigable if a leaf can float down it."

The ten-year court battle has cost the Johnson family more than $1 million. No longer able to afford lawyers, they were eventually forced to represent themselves. As a result, they were unable to mount an effective defense, and in January 2005 a district court judge sided with the EPA and ordered the family to pay a $75,000 fine and restore the site—estimated to cost $1.1 million.

The Pacific Legal Foundation (PLF) has stepped in to represent the Johnsons without charge. They are appealing the decision, and PLF and the Johnsons are prepared, if necessary, to take the case to the U.S. Supreme Court.

A POND FOR THE GREATER GOOD
(Snohomish, Washington)

A county's wetlands regulation required Arnold and Marilyn Hansen to fix the drainage problems for surrounding properties. It cost them dearly.

In 1997 the couple purchased four acres with the intent to develop them into a commercial use. After seven years, it looked like a great investment—the value of the land had increased to $280,000.

Although the Hansens obtained a permit to grade and

fill a low-lying area on the site, they decided to wait until dirt was available from the construction of a proposed highway adjacent to their land. It was a decision that they would regret.

Over the years, the low spot on the land had received runoff from other properties in the area. Snohomish County, therefore, decided that 96 percent of the Hansens' property should serve as a permanent retention pond. Of the 170,000 square feet of the land they owned, only 7,200 square feet would be usable.

The county informed the couple that they could build only one single-family home on the property. The fact that there were several commercial businesses next to the property—a diesel truck repair facility and a farm equipment dealer—and that the land fronted a major highway made it an unlikely location for a private residence.

The Hansens filed suit against the county, seeking compensation. Although the couple provided undisputed expert testimony that the construction cost of the house would exceed what it could be sold for, the county claimed that their property was still economically usable and did not constitute a taking.

After ten years of legal wrangling, a district court judge ruled in favor of the county. Eventually an appeals court dismissed the case, citing a technicality—the Hansens had never applied for a building permit.

Part IV

ENDANGERED SPECIES

18

THE ENDANGERED SPECIES ACT GONE WILD!

"The Endangered Species Act is one of the worst laws that has ever been passed in American history. It is bad for people and it is bad for species."

—R. J. SMITH, SENIOR FELLOW,
COMPETITIVE ENTERPRISE INSTITUTE AND THE
NATIONAL CENTER FOR PUBLIC POLICY RESEARCH

No one would deny that we must reasonably protect animals and plants from extinction. But the federal law designed to protect species is being abused by bureaucrats, environmental extremists, and NIMBY (Not In My Back Yard) groups who seek to undermine our fundamental private property rights.

Are we facing an endangered species extinction crisis? Environmentalists like William Snape believe so.

Snape, the chairman of the Endangered Species Coalition, says that "species and the habitat they depend upon, are declining by the day" and that scientists have concluded that "we are facing the sixth extinction spasm that our planet has ever known."

Patrick Moore, the cofounder of Greenpeace, disagrees. Moore, who left the group in 1986 because he said its focus was anti-capitalism, says the biodiversity on earth today is higher now than at any time during the 3.5 million years of life on our planet. "There is no mass extinction of species taking place on earth today," he says. "It is a complete fabrication."

Nevertheless, the U.S. Fish and Wildlife Service (US-FWS) and other supporters of the Endangered Species Act of 1973 (ESA) claim an estimated 227 animal and plant species would have become extinct between 1973 and 2004 without the law. One of the more notable, and most recent, success stories they cite is that of the bald eagle.

The USFWS notes that a survey in 1963 taken by the National Audubon Society estimated 487 active nesting pairs of bald eagles in the lower forty-eight states. Their decline began in the nineteenth century as a result of forest clearing, loss of other habitat due to development, and widespread shooting.

Beginning in the mid-1960s, scientific testing confirmed that the pesticide DDT and related chemicals were found to contaminate the fish eaten by us and the eagle. Research indicated that the pesticides impaired the birds' calcium levels resulting in egg shells that were too thin to withstand the weight of incubating parents.

Today, it is estimated that there are more than 9,700 breeding pairs of bald eagles in the lower forty-eight states and their numbers continue to rise. Environmental organizations attribute the bird's removal, or "delisting," from the endangered species list to ESA regulation. Critics of the act argue, however, that the banning of DDT by the Environmental Protection Agency in 1972 was the primary reason that the number of eagles has increased—not legal protections under the act.

The campaign to protect the endangered gray whale, which popularized the bumper sticker "Save the Whales," is another case in point. The gray whale, which was near extinction sixty years ago due to commercial whaling, was originally protected under the Marine Mammal Protection

Act of 1972. It was then listed as an endangered species under the ESA.

Today, between 15,000 and 18,000 whales live in the eastern Pacific Ocean. Again, environmentalists maintain the act was the chief contributor to the increase in the whales' population. However, ESA regulations had little do with it since whalers stopped harvesting gray whales in the 1940s—and their numbers have increased as a consequence.

IS THE ESA WORKING?

The claims and counterclaims of saving species from extinction and other reported recovery success stories brings into question the effectiveness of the act.

A study conducted by the Heritage Foundation, a Washington think tank, indicates that more than 1,300 animal and plant species have been listed since 1973. Sixty species have been targeted for delisting and only twenty-seven (not counting the bald eagle or the Yellowstone grizzly) have been delisted. The foundation's research claims that none of those species "recovered because of positive actions instituted by the federal government under the ESA." The study cites National Wilderness Institute statistics to prove its assertion:

- Seven species were delisted because they became extinct;
- Sixteen species were delisted due to errors in data;
- The Arctic peregrine falcon was delisted because it recovered after the DDT ban; and
- Three types of kangaroo were delisted as a "response to Australian policies."

A National Environmental Trust (NET) report recommends that rather than looking at the number of species recovered, "a more scientifically sound way to assess the effect of the law is to look at the extent to which it has prevented plants and animals from becoming extinct." Detractors of the ESA have reason to believe that many of the listed species were never in danger of extinction.

Whichever side one takes in the controversy as to whether the ESA is working or not, no one argues that the original intent of the act—to provide federal protection of species—is an admirable goal. However, the bureaucratic process, coupled with how the regulations are applied, continues to foster divisiveness between property owners and environmentalists.

THE REGULATORY MAZE

In order to develop land impacted by a listed species, owners (who need to obtain federal and state permits) must go through a lengthy and expensive permitting process. Under the law, *every* federal agency must examine whether any action it proposes might jeopardize the overall existence of a species—not individuals—or destroy or modify its habitat.

The issue of whether or not to issue these permits centers on the definition of "take." The ESA states it means "to harass, harm, pursue, hunt, shoot, wound, kill, trap, capture, collect, or attempt to engage in any such conduct." However, the Code of Federal Regulations goes one step further. It states that "harm" includes "significant habitat modification or degradation where it actually kills or injures wildlife." So not only must the species itself be protected but also the area in which the species lives.

This definition, of course, has wide-ranging implications. The habitat of a species could be one acre, one hundred acres, or much more. In the case of the Northern spotted owl, for example, its foraging area can cover up to four square miles.

Obviously, designating habitat to be protected could greatly impact the ability of property owners to use their land. Not surprisingly, the debate has led to several landmark court decisions.

In a 1995 Supreme Court case, *Babbitt v. Sweet Home Chapter of Communities for a Great Oregon*, a group of forest products industries, and others, challenged the addition of habitat to the regulations. Since the group's logging activities were in an area known to have red-cockaded woodpeckers and spotted owls, both listed species, its operations could be halted.

Sweet Home claimed that when Congress passed the ESA, it did not intend the word "take" to include habitat. The district court disagreed. It granted what is known as "summary judgment" and dismissed the case without going to trial. *Sweet Home* appealed.

The court of appeals reversed the lower court's ruling and concluded that the word "harm," like other words in the definition of "take," should be read to mean a perpetrator's direct use of force against the animal. The government appealed to the Supreme Court.

In a 6-to-3 decision, the high court concluded that the government did not exceed its authority when including habitat. The majority opinion stated that "harm" includes "indirect as well as direct injuries." It added that including habitat protection was reasonable, given "the ESA's broad purpose of providing comprehensive protection for endangered and threatened species."

Justice Scalia wrote a dissenting opinion and was joined by Chief Justice Rehnquist and Justice Thomas. Scalia argued that the Court's opinion "imposed unfairness to the point of financial ruin—not just upon the rich, but upon the simplest farmer who finds his land [offered for] national zoological use."

STRETCHING THE CONSTITUTION

Endangered species law then took a bizarre twist. Cases citing the Commerce Clause in the Constitution to protect a species began to surface.

Many constitutional scholars today argue that the broad interpretation of the Commerce Clause allows the federal government to involve itself in essentially every aspect of society. But how can Article I, section 8, of the Constitution, which gives Congress the power "to regulate commerce with foreign nations and among the several States" be applied to protecting endangered or threatened species?

The 1997 case known as *National Association of Home Builders (NAHB) v. Norton* involved the Delhi Sands Flower-Loving Fly in California. The U.S. Fish and Wildlife Service declared the fly an endangered species one day before the groundbreaking of a regional medical center, where, it was estimated, there were between six to eight colonies of the fly.

The NAHB sued the U.S. Department of the Interior. The government and its allies alleged that the Commerce Clause (specifically the powers to regulate interstate commerce) protects the fly. But scientific evidence showed that the fly is found only in a small part of southern California. In addition, the flies have no commercial value—they are neither bought nor sold.

Despite these facts, the Court of Appeals of the District of Columbia (considered to be the second highest court in the land) ruled that the flies could be protected under the ESA. The NAHB and many in the legal community were bewildered with the court's logic, or lack thereof.

A case in 1999 known as *Gibbs v. Babbitt* involved two North Carolina ranchers whose livestock—as well as the child of one rancher—were threatened by red wolves. The wolves were listed as an endangered species and in 1986 were reintroduced onto a national wildlife refuge in western North Carolina by the USFWS in an attempt to increase their numbers. The program then expanded in 1993 to include the release of wolves within a national refuge in Tennessee.

Because of the success of the programs, the wolves wandered across state lines and onto private property. In October 1990 a rancher shot a wolf that he feared threatened his cattle. The federal government prosecuted him and he pleaded guilty. But the case triggered opposition to the red wolf program and questioned whether the government had the authority to stop property owners from shooting wolves that threatened them or their livestock.

Ranchers filed suit in District Court. The court sided with the government's position that the Commerce Clause gave it the power to regulate the activity on private land. The court found that red wolves are "things of commerce" because they have moved across state lines and their movement is followed by "tourists, academics, and scientists." They also found that the tourism the wolves generate affects interstate commerce. The stunned ranchers appealed to the Fourth District Court of Appeals.

The appeals court agreed with the lower court's rationale. In fact, it wrote that scientific evidence *might* one day find a commercial use for wolves, or that trade in wolf pelts *might* one day be reestablished. The ranchers appealed the ruling to the U.S. Supreme Court, but the Court declined to hear the case.

MORE COMMERCE CLAUSE MADNESS

The 2003 *Rancho Viejo v. Norton* also dealt with the Commerce Clause. A developer was unable to obtain permits necessary to start construction of a housing project in San

Diego County, California, when the USFWS asserted that a portion of the land had *potential* habitat for the endangered Arroyo toad. The service also threatened prosecution for the erection of a fence that allegedly interfered with the migration of the toads.

The case made it to the District of Columbia Court of Appeals. The court cited *NAHB v. Norton* (the fly case) as precedent and refused to hear the case by siding with a lower court's summary judgment ruling that gave the government authority to regulate the project. Notably, Judge Roberts, now Chief Justice of the Supreme Court, disagreed.

Also in 2003, a Texas developer, GDF Realty Investments, purchased land to build a Wal-Mart store. The USFWS found that endangered spiders and insects inhabited caves on a portion of the property. Scientific evidence proved that the species were strictly cave dwellers and did not leave the cave. In fact, one species, a mold beetle, doesn't even have eyes.

It was clear that these insects did not have any commercial value. It was also determined that they were found only in Travis County, Texas. Despite the developer's attempts to avoid the insects' habitat, the company could not satisfy the concerns of USFWS and environmental extremists. Permits to allow the development to go forward were denied and the company sued in District Court.

Once again, the court cited the Commerce Clause precedent. It granted the government summary judgment and dismissed the case. The developers appealed to the Fifth District Court of Appeals, but it sided with the lower court.

The majority opinion of the appeals court stated that the protection of the cave bugs "is integral to achieving Congress's rational purpose in enacting the ESA." One judge, Edith Jones, disagreed and argued that for the sake of one-eighth-inch long cave insects, "which lack any known value of commerce, much less interstate commerce, the panel [of judges] crafted a constitutionally limitless theory of federal protection."

Some environmental leaders recognize that there are problems with the ESA. Michael Bean, an attorney with Environmental Defense Fund, recognizes that there is "increasing evidence that some private landowners are actively managing their land so as to avoid potential endangered species." That's putting it mildly. In fact, some owners whose very livelihoods may be at stake have adopted a "shoot, shovel, and shut up" attitude.

But William Snape, Chairman of the Endangered Species Coalition, downplays the impact on property owners. He says, "There just aren't private landowners that I can identify where the value of their property has radically declined as a result of the [act]. These landowners just don't exist."

The following horror stories may convince us otherwise.

19

HELD HOSTAGE BY A FLY
(Colton, California)

"God in His wisdom made the fly. And then forgot to tell us why."

—OGDEN NASH, POET AND HUMORIST

The City of Colton, California, has been in a battle with the U.S. Fish and Wildlife Service (USFWS) over an endangered species on the brink of extinction. The agency has claimed that this species holds the key to maintaining a valuable ecosystem. The ecosystem: sand dunes. The species: the Delhi Sands flower-loving fly.

The orange-brown and black fly is the first (and only) fly to be placed on the endangered species list. Its only known habitat are sand dunes situated in Riverside and San Bernardino counties in California, and occur within an eight-to-eleven-mile radius straddling two interstate freeways, I-10 and I-60. Biologists estimate that the fly's range once covered forty square miles. As a result of intense urban development, however, only eight colonies exist on 200 acres of scattered habitat—all but ten acres private.

Considered one of the largest in the world at one inch long, the fly is capable of stationary, hovering flight like the hummingbird and feeds on the nectar of wildflowers. The flies emerge from the sand to mate for only about a week in August or September between 10:00 a.m. and 2:00 p.m. The female then deposits her eggs in the sand and both it and the male die. The fly's total life span is not known, but it has been estimated to be between two days and three weeks long.

A FLY IN THE OINTMENT

In January 1990 the City of Colton, and other surrounding communities, were aware that biological studies for the fly were being conducted. Riverside and San Bernardino counties have one of the largest number of endangered or threatened animal and plant species in the country, but no one imagined that a fly would make the list.

The emergency listing was done in dramatic fashion. Within twenty-four hours of the breaking ground of the $487 million Arrowhead Regional Medical Center in Colton, a few biologists identified eight flies buzzing around the site and informed the USFWS. The agency then placed the flies on the endangered species list and construction was halted. The uproar was immediate, particularly since the city desperately needed the jobs and an economic shot-in-the-arm.

Outraged residents brought giant flyswatters with them to public hearings to protest the listing. There were jokes about killing off the flies at night, but that would carry a possible penalty of $200,000 per fly and one year in jail.

A USFWS official, Linda Dawes, demanded that the county set aside the entire 68-acre hospital site as a preserve for the fly. In a sworn affidavit, a witness said Dawes even went so far as to demand that Interstate 10, an eight-lane, heavily traveled freeway, be shut down—or at the very least traffic slowed to 15 miles per hour—during August and September when the flies make their annual sojourn.

Dawes was concerned that the rare insects would end up on the windshield of a speeding car.

The county hospital was eventually able to move forward, but not without paying a heavy price. In addition to spending hundreds of thousands of dollars and more than a year of effort to redesign the project to accommodate the flies, the county was forced to set aside ten acres valued at $4 million for a preserve—approximately $500,000 for each fly!

MAKE WAY FOR THE FLY

Linda Dawes, who left the USFWS soon after the agency's ruling, did not manage to close Interstate 10, but it was not the end of the county's troubles. A study showed that the intersection at the hospital's main entrance would become severely congested once the facility opened its doors. Emergency access would be slowed and could ultimately cost human lives.

The county proposed to reconfigure the interchange to alleviate the congestion, but the USFWS balked and said that the improvements would "encroach" on the fly's "migration corridor," which would be a violation of the Endangered Species Act. Although no flies have ever been documented as using this corridor, the theory that it *could* was good enough for the agency.

Government biologists determined that the flies would need to travel unimpeded over vegetation from the hospital's fly preserve to another suitable habitat about a quarter-mile away. They guessed that the flies would head west, following a 100-foot-wide corridor that would be designed for them, and then make a 90-degree turn to the north where the flyway narrows to 30 feet, and then follow that path for 700 feet.

At the other end of the flyway, the flies would then make a 90-degree turn west and cross a four-lane street to reach their cozy habitat. The biologists were, of course, assuming the flies would follow the flight plan designed for them. But how would they make it across a busy street?

The county proposed an eighteen-foot-wide flying corridor to avoid oncoming traffic, but the service said the proposal was "not biologically justified" and threatened to sue if it went forward without "mitigating" for the flyway. In other words, the county had to buy more land to serve as a fly preserve.

The USFWS then found another project that jeopardized the fly. An electric transmission line and substation had to be built to serve the hospital. The most suitable engineering location was one which the USFWS claimed would be a subsequent permanent loss of 2.5 acres of fly habitat and a temporary disturbance of 2.2 acres during construction.

Since it was not economically feasible to find a new site, the City of Colton had no choice but to acquire an additional 7.5 acres, fund it with a $66,250 *endowment* to maintain and enhance the acreage, and agree to a host of other measures to protect the flies during the construction and operation of the facility.

In another instance, the city sought to clean up an area south of the interchange where garbage had piled up along both sides of a quarter-mile stretch of a two-lane road that paralleled the freeway. Living room sofas, TV sets, old clothing, and garden trimmings were among the trash items. The USFWS told the city that certain heavy equipment could not be used because it could disturb the sand dunes which served as habitat for the flies.

The city also had to deal with the USFWS in order to build a $10 million sports park just north of Interstate 10. The park would have provided much-needed recreational facilities and spurred the development of a hotel and restaurant. However, when a survey uncovered some flies, the USFWS put a stop to the plans. It did, however, agree to let the park move forward, but only after it demanded $3 million to acquire substitute land. The city declined.

Colton was not the only city to be swept up in the fly hysteria. A developer in Mira Loma that wanted to build a warehouse project that allegedly would have destroyed fly habitat was sued twice by an environmental group. In order to settle the suits, the company was forced to pay $162,500: $50,000 to build a park; $82,500 to the Center of Community Action and Environmental Justice to purchase fly habitat; and the remaining $30,000 to the center to pay legal costs.

Riverside County Supervisor John Tavaglione, who helped broker the deal, said, "That's the most ridiculous thing I have ever head of. There's no science to back up that the fly is endangered or that there is fly habitat in the area."

GOVERNMENT LAND GRAB

The Agua Mansa Enterprise Zone, a 10,000-acre, state-sponsored zone covering parts of western Riverside/San Bernardino counties, including the cities of Colton, Rialto, and Riverside, was not immune to the fly. The zone offers tax incentives and low-cost and long-term financing for heavy and medium manufacturers wanting to relocate or expand.

CanFibre, a Canadian manufacturer of recycled fiberboard, was told by the USFWS that its entire 300-acre property was occupied by the fly. In order to go forward with a proposed manufacturing plant, the company was told it would have to provide sixty-five acres to "mitigate" the problem. A biologist working for the company said the government's demands rest on "rhetorical horse manure." As in the hospital case, the feds eventually backed down—but not before they extracted $450,000 from the company.

The USFWS then saw an opportunity to grab more land in the enterprise zone without paying for it. By using a mechanism known as a "habitat conservation plan," the agency claimed that setting aside and preserving 200 to 300 acres for flies would benefit landowners in the area.

Those owners whose land was targeted for a fly preserve would have their lands taken. Other property owners in the vicinity who wanted to develop their land (even though they might not have flies on it) would be charged "mitiga-tion fees" in order to compensate their neighbors who lost their land. Surrounding property owners, therefore, would be forced to pay for the government's seizure of land at an estimated cost of $10.5 million!

The major selling point of this scheme would be to pro-vide certainty that property owners could put their lands to use but also avoid the lengthy, expensive, and uncertain permitting process. Property owners with bank debt did not have a choice.

One would think that by reading the takings clause of the Fifth Amendment "nor shall private property be taken for public use without just compensation" that the actions the USFWS took to protect the fly were not constitutional. The National Association of Homebuilders (NAHB) cer-tainly thought so.

THE ENDANGERED SPECIES ACT (ESA) UNDER ATTACK

In 1997 the NAHB sued the U.S. Department of the Inte-rior. In *NAHB v. Norton* (the Secretary of the Interior), the government and environmental groups alleged that the

Commerce Clause, specifically the powers to regulate interstate commerce, protects the fly.

The District Court of Appeals of the District of Columbia agreed and ruled that the flies could be protected under the ESA—even though research studies showed that the fly had no commercial value and was found only in Riverside and San Bernardino counties. Judge Patricia Wald, writing for the majority, said: "the prohibition of takings of endangered animals falls under Congress' authority to keep the channels of interstate commerce from immoral and injurious uses."

Since the courts were unwilling to fix the stark inequities of the act, members of Congress took a shot. In September 2004, sensing that the protection of a fly would be a prime example of how the act was being abused by bureaucrats and environmentalist extremists, the Chairman of the House Natural Resources Committee, Congressman Richard Pombo, convened a field hearing in Fontana, California.

Pombo, a California Republican and an outspoken protector of property rights (and considered an opponent of the environmental community) was joined by Congressman Joe Baca, a Democrat from Rialto. The hearing was well attended and included public officials, developers, homeowners, biologists, and environmental groups.

Several stories about how both public and private projects had been delayed or stopped by the fly were told. In addition to the Colton cases, the mayor of Fontana testified that several important freeway interchanges that were drastically needed in the area to alleviate traffic congestion had been held up for three years because they were allegedly in the fly's habitat.

At one point in the discussion, Congressman Baca asked the audience if anyone had ever seen one of these flies alive. The Executive Director for the Endangered Habitats League conceded he had only seen a dead specimen but added that he felt it was important to "preserve all of Creation—including the Delhi Sands Fly and its ecosystem." Baca then demonstrated how he would react if an unidentified fly landed in front of him, and rolled up a newspaper.

Immediately following the hearing, the Defenders of Wildlife, Earth Justice, the Sierra Club, the National Wildlife Federation, and other environmental organizations released an "Editorial Advisory." In the release, the groups attacked Pombo and Baca, saying that they were staunch opponents of wildlife conservation and that they were simply advancing "a radical agenda driven by developer and corporate interests aimed at reversing thirty years of endangered species protection in America." The field hearings were called "a farce," "a waste of taxpayers' money," and "a violation of democratic principles." The advisory encouraged environmental groups to take an editorial stance against any bills intended to reform the Endangered Species Act.

A year later, the U.S. House of Representatives passed the Threatened and Endangered Species Recovery Act (H. R. 3824) in a 229-to-193 vote. A Senate bill, however, failed to get support. In November 2006 Representative Pombo lost his bid for re-election due, in large part, to the well-funded and concerted efforts by national environmental groups.

The one-inch-long fly continues to be one of the more extreme examples of how the ESA has been used to take private property without compensation *and* to exact money and land in return for the right to develop. Two USFWS wildlife biologists wrote an article that said that "some people will not miss the Delhi Sands flower-loving fly, but they may very well miss the other plants and animals that depend upon the important ecological role this unique insect plays."

Tell that to Colton city officials, who, after a thirteen-year battle, claim that the fly has caused them to lose at least $175 million in economic development opportunities. And also to all those people with giant fly-swatters.

20
THE CAVE BUGS ARE SAFE
(Austin, Texas)

"I don't want to negate the Endangered Species Act. I just want to sell [my] land to the government or develop it. And the way the law is now, I can't do either."
—FRED PURCELL, AUSTIN, TEXAS, DEVELOPER

Soon after Fred Purcell bought 216 acres northwest of Austin, he received a phone call that forever altered his life. The caller told him that members of EarthFirst!, an environmental extremist group linked with hundreds of eco-terrorist acts throughout the country, were huddled near the narrow entrance of a cave on his land and staging a sit-in before television news crews. Why? To protect bugs!

The scene was surreal. At one point, while the protestors ate pizza, a sheriff's deputy had to crawl into the narrow cave to arrest a protestor who had wedged himself into a crevice. Eventually, the deputy was able to pull the man out and handcuff him.

Fred, a dentist, his brother Gary, and other partners had formed investment partnerships to buy the land for nearly $3 million in 1983. It lay at the intersection of two highways in an area of Austin known as Four Corners, one of the fastest-growing areas in the country. It was also located in the Balcones Canyonlands, tracts of land designated for a future preserve area.

The proposed 30,000-area preserve comprised a series of underground springs that were known to percolate through limestone rock and form caves, sinkholes, and steep canyons. It was also home to two endangered bird species—the Golden-cheeked warbler and the Black-capped vireo—as well as certain cave bugs.

JUSTICE AND THE BUGS ARE BLIND

Prior to the EarthFirst! "cave-in" (and despite the attempts of environmental groups to stop the proposed development), Fred and his partners had obtained final development approval from the City of Austin. The investment group spent more than $10 million to construct water and sewer lines and other utility improvements, with the intent to sell parcels for a proposed Wal-Mart, apartments, and office space. The U.S. Fish and Wildlife Service (USFWS) and environmentalists, however, had a different plan.

In 1988 the USFWS placed six separate cave invertebrates that were thought to be found on the property on the Endangered Species list. These spiders and beetles were all smaller than a thumbnail and specially adapted to live underground, which explains why four of the six species don't even have eyes.

Not one of these invertebrates has ever been known to leave the cave or sinkhole in which they are found. They are completely isolated in caves, and no more than a handful of naturalists—seeking them out in their habitat—have even seen the tiny creatures.

Fred Purcell and his group were well aware of the horror stories property owners had experienced once listed species are found on their property. They knew that to illegally "take" the bugs could not only slow or stop the proposed development but also subject them to hefty fines and even imprisonment.

Fred and his partners sought and followed the USFWS's advice and conducted a series of field surveys to determine the best ways to protect the bugs while developing the property.

In 1990, based upon the recommendations of the USFWS, the investment group placed gates across the entrances to the caves. In order to provide "buffer zones" for the caves, other sections of the property were also restricted from development. In addition, the group donated six acres to a nonprofit organization dedicated to environmental research. Fred and his partners could now begin to sell portions of the remaining property—so they thought.

In 1991 the investment group entered into a sales contract with Inland Laboratories to develop and sell ten acres. One condition of the agreement was for Fred and his partners to provide assurance to Inland that the construction of its facilities would not violate the Endangered Species Act (ESA).

Fred Purcell requested a letter from the USFWS. Despite the partnership's having complied with all the recommendations of the USFWS, the agency refused to issue a letter. As a result, Inland terminated its contract.

Soon thereafter, Fred Purcell attempted to clear brush and trash off the property. The USFWS filed a criminal prosecution alleging that he had violated the ESA. He was convicted on two counts, but the convictions were overturned within a year.

JUST SAY NO

Over the next few years, Fred and his partners continued to propose a series of less intensive development plans to appease the agency. In each instance, however, the USFWS changed its position and made additional demands to set aside more acreage from development. During these lengthy negotiations, two more businesses that wanted to purchase large portions of the property backed out of contracts.

In order to develop *any* portion of the property, Purcell and his partners were then forced to apply for a permit under section 10 of the ESA in December 1997. These permits, which can cost several hundred thousand dollars and take years to obtain, require the property owner to develop a plan to protect habitat of the species, known as a Habitat Conservation Plan (HCP). Theoretically, HCPs protect the areas where species thrive while at the same time allowing for reasonable development.

It is not uncommon, however, that when proposals are submitted to the USFWS, they are deemed inadequate. In many cases, the agency recommends more property be excluded from development and kept in its natural state— without compensation. Purcell's application proved to be no exception.

The investment group filed seven separate applications covering different sections of the property. After the agency failed to act on the applications in a timely manner, Purcell and his lawyers asked for written denials. The investment group, according to later court documents, alleged that the USFWS refused to process the paperwork in order to prevent the group from challenging the agency's actions. The partnership had no choice but to sue.

The agency answered the complaint by saying that the applications were "deficient" and that the proposed project would be an illegal "take" of the bugs. The agency also stated that they had adopted new, *unpublished* guidelines related to protecting the cave insects. Under these new "rules," the agency imposed conditions which would, in effect, prohibit development on nearly all of the land. The stalling continued until June 1999, when the court issued a final order that the USFWS had, in effect, denied the investment group's applications. This would clear the way for the group to file a lawsuit.

Deprived of the ability to develop any of the property to pay off a $4.8 million loan and pay property taxes, the group was forced to file for bankruptcy in 2000. The group would eventually lose all but seventy acres of the original 216 acres.

Fred Purcell and his partners felt they had a good "takings" case, but Paul Terrill, their Austin attorney, knew that winning a takings claim by citing the Fifth Amendment of the Constitution was a long shot. Terrill, therefore, turned to Dan Byfield with the American Land Foundation (ALF), a Texas-based property rights organization.

Terrill knew that Byfield had been looking for an endangered species case to litigate. After reviewing the case, ALF agreed to fund a lawsuit that would attempt to change the legal precedent of using the Commerce Clause—the authority Congress had under the Constitution to regulate interstate commerce—to take private property through the Endangered Species Act.

On June 15, 2000, *GDF Realty Investments Ltd. v. Babbitt* was filed in the U.S. District Court in Austin.

The government's position was that the Commerce Clause gave it the legal authority under the ESA to regulate any activity that could impact the insects. GDF argued that the Commerce Clause was not applicable: the bugs were not bought and sold and could not cross state lines.

The U.S District Court ruled against GDF. It cited that the *development* had "a substantial effect on interstate commerce." The legal community was bewildered. Terrill said that the "basic flaw in the [judge's] analysis is that the take provision in the ESA does not regulate commercial development"—it regulates the "takes" of endangered species.

PROTECTING THE WEB OF LIFE

GDF appealed to the Fifth District Court of Appeals in New Orleans. The stakes were very high for the environmental community—especially if the lower court ruling was overturned based upon the fact the insects did not cross state lines. Environmentalists believed such a decision would set a precedent that could strip protection from more than 600 listed species that were isolated to a specific state.

Lawyers for the USFWS continued to argue that the commercial activity caused by the development would be covered under the Commerce Clause and, thus, authorize the agency to halt development because it could harm both the bugs and their habitat. The judges disagreed.

The USFWS also argued that scientific interest in the insects created a substantial impact on interstate commerce since scientists would travel to Texas to study them. The court did not buy the argument.

Government lawyers further claimed that the bugs served a commercial purpose since research into endangered species has been used in the treatment of diseases. The court did not buy that argument either.

Finally, government lawyers admitted that taking the insects "does not substantially affect interstate commerce." However, their contention was that "piecemeal extinctions" of the bugs could threaten the "interdependent web" of all species and, therefore, "undercut the ESA scheme." The court agreed.

The court's majority opinion said that, according to Congress, the "essential purpose" of the ESA is to "protect the ecosystems upon which we and other species depend." It then went on to say that the ESA is "an economic regulatory scheme" and that "the regulation of the intrastate takes of the cave species is an essential part of it."

Despite the crushing loss, GDF and their attorneys were

emboldened by the dissenting opinion of six judges on the panel. They said that the "commerce clause regulates commerce, not ecosystems." Therefore, in August 2004, GDF appealed to the U.S. Supreme Court.

JUST TAKE THE CASE

The environmental community was, of course, staunchly opposed to the high court hearing the case. John Kostyack, a senior counsel at the National Wildlife Federation, said the challenge was a desperate "Hail Mary" attempt to resurrect an old argument against the law. He noted, however, that an adverse ruling in the case "could gut the law" and that hundreds of species that do not cross state borders "would be headed quickly for extinction."

Among the many briefs filed on behalf of GDF (including one by the State of Texas), the Washington Legal Foundation urged the Court to review and reverse the Fifth Circuit's opinion. It said the ruling, in effect, gives the US-FWS "essentially unlimited authority to regulate local land development across the country in the name of protecting endangered species."

On June 14, 2005, however, the Supreme Court denied GDF's request. Fred Purcell and his group's seventeen-year battle was over. John Kostyack, of the National Wildlife Federation, said the Court's decision served as a reminder that "the protection of the endangered species is a national priority" and that as a country "we have an interest in the protection of these natural systems that we rely upon for our quality of life." If that is the case, Fred Purcell and his partners believe that the public should then buy their land.

"Despite what the Constitution says," says Dr. Purcell, "no one is ever going to pay you. That is reality." He adds,

"If you are unfortunate enough to wake up one morning and find out you've got an endangered species on your property, it is your burden."

Environmentalists, however, can go to sleep every night with the comfort of knowing that the Endangered Species Act—and the cave bugs—are safe.

A GRIZZLY LESSON TO LEARN
(Dupuyer, Montana)

"No two bears will do the same thing in a given situation, and a bear may not do the same thing twice."
—LARRY KANIUT, AUTHOR OF *ALASKA BEAR TALES*

On a snowy September night in 1989, Montana sheep rancher John Shuler and his wife were awakened at 10:30 p.m. by the sound of crunching bones. The grizzly bears had returned.

John jumped out of bed. He looked out the window toward his illuminated sheep pen and noticed that the sheep were circling nervously. He then saw something move along the fence line heading toward the pen.

John grabbed his flashlight and his rifle. He ran outside onto the porch barefoot and in his underwear. His dog, Boone, was barking. He noticed the sheep scatter and then start piling up on one another.

John ran toward the pen. He climbed the fence and headed toward the center of it. Snow began to fall and made it difficult to see.

The sheep were boiling all around him and then, out of the darkness three grizzlies sprinted past him. Startled, he dropped his flashlight and fired at the bears.

Suddenly, a fourth bear rose on its hind legs within

thirty feet of him. He fired at the bear's throat. It fell to the ground and let out a vicious roar. It then got up, but John lost sight of it amid the heavy snowfall.

He rushed behind an A-frame structure into which lambs go to be protected from the wind. He then heard the gate rattle at the far end of the pen but couldn't tell if it was sheep hitting the fence or the bear climbing over it.

After the fence stopped rattling and when he thought it was safe, John retrieved his flashlight and then stood where he had fired the shot. There, in the snow, were blood stains.

He followed the trail of blood to where the fence had rattled and then picked up the trails of the three other grizzlies. Deciding that the bears had been scared off, John went back to the house.

After getting dressed, he jumped into his pickup truck to look for the bear's carcass. Visibility got so bad, however, that he abandoned his search.

Fearing the bears would return and kill his sheep, John spent the rest of the night awake sitting with the rifle on his lap and Boone by his side. He wondered whether he had killed the grizzly or simply wounded it. He knew a wounded bear could be very dangerous and posed a threat to his neighbors.

A LIFE AND DEATH ENCOUNTER

At first light the next morning, John and Boone climbed into the pickup truck to resume the search. He drove south toward a creek where he figured the bear had headed.

He stopped in an open pasture. As soon as he let Boone out, the dog disappeared into a low marshy area.

John, armed with his rifle, got out of the truck and walked toward the creek. He then noticed that Boone had stopped and was in a pointing position. John walked toward him with his rifle raised, but there was no sign of the bear.

He turned back toward the truck and locked glances with a bear sitting on its haunches. The bear quickly rose and charged.

John raised his rifle. He fired but missed. He reloaded. The bear stumbled and fell within one hundred feet of him. It got up and charged. John fired again. The bear fell.

The animal was badly wounded and clawed its way into a stand of trees and then stopped. John cautiously approached it. The female bear was barely breathing. He ended her suffering.

When John got home, he received a call from a wildlife officer who inquired if they were "having any problems" with grizzlies. John told the officer what had happened.

A few days later, officers came to the ranch. They retrieved the carcass and interviewed John at his home. He claimed that when the bear charged him it was limping and he figured it had been the one he had shot the night before.

John knew it was against the law to kill a grizzly, listed as a "threatened species" under the Endangered Species Act (ESA), unless it was in self-defense. Although he did not specifically claim he killed the bear in self-defense, he figured the details of the encounter made it apparent. He also told the USFWS officers about his other encounters with bears (all of which had been reported to the service at the time) and the fact that he had lost eleven sheep.

After the officers left, John believed that was the end of the matter, but he was badly mistaken. The U.S. Fish and Wildlife Service (USFWS) charged John with violating the ESA, claiming he was not acting in self-defense, and sought a fine of $7,000.

Despite John's story that the grizzly charged him, the

USFWS can legally maintain that charging is not necessarily a sign of an "imminent attack" but can, instead, merely be a "false charge." Considering that the grizzly bear is one of North America's most dangerous predators, who is going to take a chance that a bear is just trying to make you run for your life?

BEAR FACTS

There is no dispute that the grizzly, whose Latin name is *Ursus arctos horribilis*, can be ferocious toward animals and humans alike. An adult male stands seven feet and weighs 300 to 600 pounds (occasionally more than 800) and females weigh between 200 to 400 pounds. Despite their size, the bears can run as fast as 40 miles per hour.

Grizzlies are also known to have very keen senses of hearing and smell. Their teeth are large, and although they are not particularly sharp, the beast's powerful jaw muscles allow it to penetrate deep into soft tissue and crush bones with ease. As a result, the injuries inflicted on a human are a series of punctures, tearing and shredding.

The claws on the bear's front pads can be as long as human fingers and can cause deep gashes. One powerful swipe of its paw, especially to the head or neck, can easily kill someone.

In addition to the physical attributes of the grizzlies, they are extremely unpredictable. In many reported cases, a grizzly attacks and backs off, sometimes waiting and watching, and then resumes mauling the victim— sometimes going for the head if the victim moves. They have also been known to swipe their claws across the genital areas to test for signs of life.

Some experts advise to play dead and not move. However, if a bear is hungry, angered, or if a person is unfortunate enough to be between it and a cub, one may not have time to play dead.

The use of pepper spray has also been known to deter a bear from attacking. The cloud of spray startles the bear and can put it in a defensive mode. Once it has breathed the chemical, its sense of smell is instantly shut down, which confuses it. The spray is also an inflammatory agent. It irritates a bear's mucous membranes and can cause choking and difficulty breathing. In some cases it can also cause temporary blindness.

Another compelling reason to use pepper spray is that sometimes charges by grizzlies are not "attacks" but are simply attempts to discourage and intimidate intruders. Trying to read a bear's behavior, however, can be tricky, especially when someone is on the verge of panic. But how does one know when the bear is bluffing or not?

Some outdoorsmen believe a bullet is the best defense. A bullet, however, is not much wider than a pencil and to stop a rampaging bear the bullet would have to hit a vital area such as its face or upper chest. Since grizzlies lower their heads when attacking, some hunting guides suggest aiming for its snout—a high shot goes into the upper skull or even over the top to the neck or spine. A low shot could hit the throat, chest, or even a shoulder or leg which *could* stop the animal.

Whether someone uses pepper spray or a bullet, it is hard for someone under attack to second-guess the actions of a bear. In John Shuler's case, it was self-defense. Especially after wounding the bear, he had no choice but to kill it.

A LEGAL QUAGMIRE

William Perry Pendley of Mountain States Legal Foundation, a nonprofit public interest law firm based in Lakewood, Colorado, had read about John Shuler's incident and offered to represent the sheep rancher at no cost. Shuler (in the company of Pendley) told his story to an administrative judge in Great Falls, Montana.

During the proceedings, a game warden testified that Shuler "said he was glad he had killed the grizzly." This same warden had worked with Shuler to install an electric fence around the sheep pen after Shuler's run-in with the bears. The warden's testimony, however, did not persuade the judge who—months later—ruled that John Shuler was in reasonable fear for his life on the night of the incident and that he acted in self-defense. Shockingly, the judge also ruled that as a "matter of law," Shuler had no right to go outside his house that night. By doing so, he was placing himself in the "zone of imminent danger."

The fine was reduced from $7,000 to $4,000. In an effort to clear his name, however, Shuler and his lawyers decided to appeal to the Ad Hoc Board of Appeals with the Department of the Interior.

Briefs were sent to the two-member panel. It took them three years to issue a ruling. They said that John had the right to arm himself and go outside. Nevertheless, the panel ruled that he could not claim self-defense since he took his dog with him when he killed the grizzly. The panel's claim: the dog had provoked the grizzly.

The law was clear, the panel said, that if anyone provokes a situation they may not claim self-defense. John was held liable for violating the ESA and the fine was raised from $4,000 to $5,000.

One of the more astounding conclusions the panel reached was that when the bear rose up and growled, it could have been nothing more than a greeting. A wounded, angry bear greeting him?

Shuler filed a lawsuit in Montana federal court to challenge the fine. Based upon the factual record in the two previous cases, the court ruled in favor of Shuler—that he had acted in self-defense and not violated the ESA.

After an eight-year battle, which cost Mountain States Legal Foundation more than $225,000, Shuler and his wife could now rest easy. However, one issue remained—the reimbursement for legal fees.

Federal law authorizes a court to order a federal agency to pay a person's attorney fees when the government's position was "not substantially justified."

Mountain States' attorney William Perry Pendley requested that Mountain States be paid its legal fees. The court, however, said the USFWS was right to prosecute Shuler and that the attorneys were not entitled to *any* fees.

In November 2000 the lawyers appealed to the U.S. Court of Appeals for the Ninth Circuit in San Francisco—a highly controversial, liberal-leaning court. The Ninth Circuit ruled that "there was substantial evidence to support the theory that Shuler wrongfully killed the grizzly bear in question, and even that he was pleased to have had the opportunity to carry out the deed." Reimbursement of the attorney's fees was denied.

A DETERRENT?

Although Shuler was able to clear his name, the court case was far from a defeat for the USFWS. The agency clearly demonstrated its willingness to aggressively prosecute anyone who killed a grizzly bear. It also sent a very strong message that such a prosecution would be extremely time-consuming and expensive. One consequence, however, is how John Shuler's case influences what some people may do when confronted by a grizzly.

In October 1998 a grizzly charged Wyoming hunter Pat vanVleet. Although armed with a rifle, he laid his rifle down. He took a can of pepper spray from his belt and sprayed the rampaging bear.

The grizzly stopped. It took a whiff and then continued its charge. The bear bit into vanVleet's stomach, caught onto his large belt buckle, and then flung him around.

A shot rang out. The grizzly fell dead. Thanks to a hunting partner's lucky, life-saving shot, vanVleet survived, but he was badly injured.

The attack made news. From his hospital bed, vanVleet told reporters the reason he did not shoot the bear was because he didn't want to lose his hunting license. A few months later, he also told Mountain States' attorney William Perry Pendley that he had heard about John Shuler's story and did not want to go through the same ordeal.

It appears some people are more afraid of getting mauled by the government than by the mighty grizzly.

22

STORY SNAPSHOTS

THE COURTS DON'T GIVE A HOOT
(Linn County, Oregon)

A pair of Northern spotted owls, a "threatened species," stopped Alvin and Marsha Seiber from harvesting trees on their land—despite the fact that the birds were not even located on their land.

Alvin, a retired postal worker, and his wife, owned 200 acres and relied on the income from logging operations for their retirement. A pair of owls, known as "Little Wiley" and his mate, however, had been identified in the vicinity and (at one point) had nested there.

The Seibers applied for a permit through the U.S. Fish and Wildlife Service (USFWS) in 1994 to allow them to harvest the trees. Despite the couple's willingness to cut the trees outside the owl's breeding season and to replant within its habitat, the permit was denied.

They then applied for a permit with the Oregon Department of Forestry. The department said that forty acres of their land had to be set aside from harvesting since it was within a seventy-acre "protection zone" for the owl pair. The couple filed suit in federal court alleging that their property had been taken and sought compensation.

Soon after the suit was filed, the USFWS reported that the pair of owls no longer occupied the area and that logging on the forty acres could proceed. That, however, was not the end of the story.

A flurry of legal challenges followed. The Seibers alleged that there had been a "temporary taking" of their property (from 1994 to 2000) and that they should be compensated for their loss of income from the forty acres—estimated to be $300,000.

The government, with the support of several environmental organizations, however, cited legal precedent to claim that the Seibers should be denied compensation for the taking. It cited a case that said "regulatory restrictions . . . do not constitute physical takings." In addition, the government cited a case that set a standard that since the *whole* parcel (200 acres) was not impacted by regulations, then the Seibers were not denied "all economically beneficial or productive use" and, therefore, were not entitled to any compensation. The Oregon Supreme Court agreed.

During the four years of litigation, the couple had paid property taxes on property they could not use. Much of the timber on the forty acres had also blown down because of root rot disease and could not be sold.

A DIFFERENT KIND OF BEETLEMANIA
(Lusby, Maryland)

Richard Bannister and his wife had to act fast. Their new Chesapeake Bay home was about to slide down a sixty-foot cliff due to erosion problems.

The couple gained approval from the U.S. Army Corps of Engineers to build a stone retaining wall that sloped at the base of the cliff. The state's Natural Heritage Foundation (NHF), however, told the Bannisters that the project could harm the habitat of the rare Puritan tiger beetle.

The beetle (which is no bigger than a thumbnail) is

found along sandy shorelines and in only two places in the world: the Connecticut River in New England and a small section of the Chesapeake Bay. Although it lives twenty-two months of the year underground, the speedy little beetle— a voracious hunter—runs up and down the beach preying on other insects. Researchers have documented the mating habits of the beetle. Even though a male need only copulate with a female for a few minutes, it rides the back of the female for up to six hours.

The Bannisters' frantic calls to the NHF's state office went unreturned. A reporter with the *Calvert County Independent* newspaper, however, spoke to an official who said, "We don't have to justify everything we need to protect."

The Bannisters then wrote to the governor. Eventually, the secretary of the Department of Natural Resources proposed an alternative that was not acceptable to the Army Corps of Engineers. While the wrangling continued, a fifteen-foot section of the Bannisters' property collapsed, leaving only fourteen feet between their house and the cliff.

After nearly three years of bureaucratic bickering, a compromise was made: the Banisters had to pay to relocate the beetles before the wall could be built. One "bug-hugger," however, was quoted as saying that the consequence was that "beetle numbers in that area declined" and that "it's never recovered, and it probably will never recover."

Biologist Barry Knisley, a noted expert on the beetle, who admits the Puritan tiger beetle is not a "charismatic species," is often asked, "Why should anyone care about these insects?" He responds by saying that all forms of life should be respected and that studying the natural world in the tradition of Charles Darwin can lead to great discoveries. "It's kind of a quest for information."

Darwin also believed that the extinction of a species is a natural process.

A SECURE BORDER IS FOR THE BIRDS
(San Diego, California)

Preventing illegal immigration across our border with
Mexico is one of the most important topics of our time.
One proposed solution to stem the flow of immigrants is to
build border fences. Some environmental groups believe,
however, that protecting some endangered birds is more
important.

Among the mandates in the 1996 immigration bill
passed by Congress was to place fence fortifications along
a fourteen-mile stretch of the border south of San Diego.
A three-and-a-half-mile portion of that barrier happens to
run through coastal wetlands and, according to a 2004 biological study by environmental groups, is home to one pair
of the Least Bell's vireos and another pair of Southwestern
willow flycatchers—both endangered species.

The Office of Homeland Security sought to fill in and
improve treacherous dirt roads within a deep half-mile
canyon that were used by illegal border crossers. The roads
had also been responsible for the deaths of three Border
Patrol agents.

Both the Sierra Club and the San Diego Audubon Society sued the agency and claimed that it had not fully dis-
closed the environmental impact on the birds. The California Coastal Commission (CCC) also sued and claimed
filling the canyon (known as Smuggler's Gulch) would
damage the Tijuana River estuary.

Despite the claim by the CCC and environmental groups
that the Office of Homeland Security did not have the authority to proceed with the project, Secretary Michael
Chertoff waived all environmental challenges. A federal
judge then agreed that Congress had delegated such authority to the agency.

"From our office's perspective," said a deputy director

of the California Office of Homeland Security, "we think homeland security is a top concern." The CCC, however, argues that border protection "must be balanced with habitat protection."

STYMIED BY SNAILS
(Kanab, Utah)

Brandt Child's 450-acre property had a spring-fed wetlands surrounded by desert in southern Utah. This oasis was located between two national parks and was ideally suited for Child's planned RV campground and golf course. The land, however, was also home to around 100,000 Kanab ambersnails.

Child's story begins in 1993. Soon after he bulldozed some brush near the lake, the U.S. Army Corps of Engineers paid him a call and threatened to fine him $25,000 for allegedly destroying the wetlands without a permit. Luckily, he was soon able to restore the damaged area and thought that was the end of it.

Child then received an unexpected visit from the U.S. Fish and Wildlife Service (USFWS) and was told that the wetlands served as habitat for the one-inch, slimy mollusk. In order to protect the creature, the service "emergency-listed" the snail as an endangered species. It then forced Child to fence the property and forbade him to work in the area. The agency also informed Child that if he failed to report a problem, he could be held accountable.

Shortly thereafter, eleven domestic geese mysteriously took up residence at the pond. Child informed the USFWS and was told that if any of the geese ate the snails, he would be fined $50,000 for each snail eaten.

The agency then ordered the Utah Department of Wildlife and Resources to shoot the geese, remove their stomachs, and forward the contents to USFWS. But when a state wildlife agent arrived, there were so many photographers and journalists present that the plan to kill the geese was abandoned. Eventually, it was determined that the geese did not harm the endangered snail.

Vicky Meretsky, a biologist with Indiana University, was delighted with the snail finding. "It was really unexpected," she said. "It gives us another population in a new location and that's always a good thing."

Child, who claims the rare slugs have cost him $2.5 million, does not quite share her enthusiasm.

THE MOUSE THAT NEVER WAS
(Chugwater, Wyoming)

All Jim and Amy LeSatz wanted to do was build their own horse-riding arena to avoid having to transport horses to a public facility twenty-five miles away. But the listing of the Preble's Meadows jumping mouse as a threatened species in 1998 put a halt to their plans.

The couple are not the only private property owners impacted by the mouse listing. Nearly 31,000 acres along streams in Colorado and Wyoming have been designated as critical habitat—including an area under rapid development. Developers and homebuilders in the area claim that the onerous regulations to protect the mouse have caused many proposed projects to be abandoned.

The mouse controversy has become so bizarre that residents of one Colorado Springs subdivision are required to keep cats on leashes for fear that mice will become their prey. In rural areas, ranchers are not permitted to clear weeds out of ditches, thereby reducing the amount of water to irrigate hayfields in the middle of summer. In addition, the mouse has also blocked the construction of badly needed reservoirs to deal with a five-year drought.

The entire scientific basis for listing the mouse, however, has been brought into question. An obscure biological report conducted more than fifty years ago that examined the skulls of three mice and skins of eleven others has been contradicted. A curator at the Denver Museum of Nature and Science has conducted DNA research that proves the mouse is actually the Bear Lodge Meadow jumping mouse—a species that is abundant and does not need protection.

Despite this evidence, the U.S. Fish and Wildlife Service has not moved to delist the mouse. An attorney for Coloradans for Water Conservation and Development that is pushing for the delisting says, "The bottom line is [critical habitat designation] has been a wonderful tool for environmental groups to try to stop things."

In the meantime, the LeSatz riding arena is on hold and developers and ranchers must continue to deal with regulations for a mouse that does not exist.

SNAKES IN THE GRASS
(Fishkill, New York)

Jay Montfort and his family-owned company wanted to expand an existing gravel mine. They battled a state conservation agency and a wealthy preservationist group over environmental issues. At the center of the fight: a den of rattlesnakes.

Since the mine's expansion would have created fifty badly needed jobs, the community supported it. The project received local approval in 1990 and then Montfort sought a mining application from New York State's Department of Environmental Conservation (DEC).

The permitting process was complicated and time-consuming. From the outset, the DEC rejected one environmental study after another. It also refused to render decisions within legally required time frames.

Permitting costs soared to more than $2 million. Five years later, after the proposal was refined to suit the agency, the plan was accepted. Scenic Hudson, a preservationist group that had bought land for preservation next to the Montfort's property, began an aggressive public relations campaign to stop the mining operation.

One alert that Scenic Hudson issued to its supporters was that "A diverse wildflower community, ideal reptile habitat, and a wilderness setting preferred by many birds hang in the balance as the mining proposal goes forward." It also challenged the plan over aesthetics and noise.

Montfort worked in good faith with Scenic Hudson to resolve their differences. Negotiations dragged on for another few years. When it appeared that Montfort's plan would be approved, two DEC employees discovered a den of Eastern timber rattlesnakes—a species listed as "threatened" by the State of New York—on the property Scenic Hudson had purchased and within 260 feet of the Montfort's existing mining operation.

In order to protect his employees, Montfort then built a four-foot, wire-mesh fence around the property. The DEC demanded that he tear it down because it would reduce the snakes' habitat. The agency also cited state law that prohibits not only killing or harming species but also lesser acts such as "disturbing, harrying, or worrying."

Montfort sued the DEC. The state's lower courts ruled against him and the Dutchess County Supreme Court dismissed the suit. Montfort could not expand his operation and no new jobs were created.

At least the snakes could live happily ever after.

THE BALD EAGLE: DELISTED IN NAME ONLY
(Sullivan Lake, Minnesota)

President Clinton announced with great fanfare on the Fourth of July 1999 that the bald eagle would be taken off the endangered species list due to its miraculous recovery. The government, however, continued to drag its feet and Edmund Contoski, a Minnesota landowner, was forced to make it live up to its commitment.

Contoski, who owns twenty-three acres of lakefront property, wanted to build five modest vacation cabins. He was planning to carve out a road until state environmental authorities identified a bald eagle's nest on his land. Even though the nest was not being used by the birds that year (they eventually returned), he was prohibited from doing any work within a 330-foot radius. "I can't even cut firewood," he said. "I can't trim a tree. I can't do anything."

Contoski, with the support of the Pacific Legal Foundation (PLF), filed a lawsuit in federal court in 2005 to demand that the bald eagle be delisted. The court ruled that the government had to remove the eagle from the "threatened" list by June 2007. That was the good news. The bad news: The "new" guidelines put in place after the delisting are essentially the same as the rules under the Endangered Species Act (ESA)—including the "no development zone" of 330 feet.

The rules, says Damien Schiff, a PLF attorney, are "as restrictive as if the ESA listing had never been lifted." Michael Bean, with the Environmental Defense Fund, said, "There is no greater tribute to the [act] than to allow its finest success story to fly off the list . . . free at last."

Edmund Contoski, however, is still not "free" to build his cabins.

23

A CALL TO ACTION

"With public sentiment, nothing can fail; without it nothing can succeed."

—**ABRAHAM LINCOLN**

Why should you care that Susette Kelo and other families had their homes seized and handed over to a private developer so that the City of New London, Connecticut, could generate more tax revenues?

Why should you care if the Stephanis brothers in Pompano Beach, Florida, had their building permits yanked by the city so that the precious ocean views of influential property owners were protected?

Why should you care if Midland, Michigan, developer John Rapanos has been threatened with a federal prison sentence and fined $13 million for moving sand around his own land?

Why should you care if the City of Colton, California, lost an estimated $175 million in badly needed economic development projects in order to protect an endangered fly?

Even though you may *think* you may never be affected by these and other types of takings, you should care:

- about simple fairness and decency for your fellow citizens;
- about the ripple effect that regulations have on the economy; and
- about how the infringement of property rights can lead to other violations of the U.S. Constitution.

Our homes, land, and businesses are an expression of ourselves. These possessions have sentimental value but also represent the fruits of our labor and serve as a family's financial security. When property owners are victimized by government officials, no-growth advocates, and environmental extremists, they experience intense feelings of personal violation—not unlike when someone is burglarized.

Obviously, the emotional strain and financial hardships these owners must endure can produce feelings of desperation, destroy families, and even cause illness. In the case of Joseph Hill, whose Massachusetts dairy farm was targeted by preservationists, the ordeal contributed to a series of heart attacks and his eventual death.

In addition to the personal toll owners face, the cost of government regulation impacts our economic prosperity. In the case of homebuilder Frank Kottschade, the City of Rochester, Minnesota, imposed so many development conditions for a proposed town house project that it increased the cost of each home by 300 percent—far beyond the reach of middle-class buyers and much higher than market prices in the area. Thousands of cases like this exist throughout the country. Is it any wonder there is a shortage of affordable housing?

The legal right of the government to seize property through eminent domain for the sake of generating tax revenues and to impose costly regulations beyond what is necessary to protect the public's health, safety, and welfare has been made possible by judges who have strayed from the intent of our Founding Fathers to respect property rights. The dangerous legal precedents they set can lead to other liberal interpretations of the Constitution and more violations. One prime example is the 1984 Supreme Court case known as *Ruckelshaus v. Monsanto*.

This case involved the Environmental Protection Agency (EPA) forcing Monsanto, an agricultural biotechnology company, to divulge trade secrets to its competitors in order to remove a barrier into the pesticide market. Utilizing legal precedent from past cases involving real estate–related takings, the Supreme Court somehow justified their unanimous ruling that the EPA's action was for a "public purpose." What's next?

You can stand by and allow the government's assault on property rights to continue—and perhaps wait until it reaches your or your neighbor's doorstep—or you can take decisive action now.

The choice is yours.

If you choose to take action:

- you must get informed and stay informed;
- you must communicate your position;
- you must band together with others;
- you must contact and engage the media;
- you must press for judicial reforms; and
- you must make politicians accountable with your vote.

This book attempts to provide a general understanding of how the government infringes on your property rights. However, in order to stay informed of eminent domain and No-Compensation and Pay-To-Play cases and issues, go to www.GovernmentPirates.com.

Among the many features on the site is a *Links & Resources* page. The nonprofit organizations listed—such as the Institute for Justice (and its sister organization Castle

Coalition), Pacific Legal Foundation, Mountain States Legal Foundation, and the American Land Foundation and others—provide valuable up-to-date information on everything from current cases, the status of ongoing litigation, reports, newsletters, and reforms. (See Appendix: Legistrative Reform Strategies on page 223.) These dedicated groups want to help you. In some instances, they will even represent property owners in court at no cost.

Effective communication is one of the keys to making politicians and bureaucrats aware of an issue. Calling the offices of elected and appointed officials is helpful, but e-mails and, more important, short, neatly written and courteous letters are the most effective.

Any attempt to reform property rights—like any other movement—also depends upon your aligning yourself with others who want to effect change. Well-organized groups that attend and speak at public hearings and town hall meetings, before city councils, state legislatures, or Congress will have a definite impact. If politicians and bureaucrats are convinced that a groundswell of support is building for a certain issue, they will be forced to address it. Holding protests (even if small) at these events can also be very effective.

Your local media can play an important role. Writing letters to the editor or submitting guest commentaries to your local newspaper can help galvanize more public support for your issue. Call and e-mail reporters who cover city council meetings. Contact radio talk shows and request to be interviewed or be a caller during a show. Also keep in mind that television news producers and investigative reporters are always looking for "ratings-rich" stories.

On the national front, Fox News Channel's *Hannity & Colmes*, CBS's *60 Minutes*, *20/20* on ABC, and other programs, in addition to nationally syndicated radio shows, have shed light on eminent domain abuse and other takings. This media attention has helped homeowners and small business owners in their fight to protect their property from the government.

In order for any media coverage to shape national public sentiment on property rights, however, there must be more reporting than what we have seen in the past. Producers and editors will only run compelling stories if they are convinced that there is a genuine public interest and, of course, that coverage will boost ratings or increase circulation. The "Tell Us Your Stories" section of Government Pirates.com was developed for that purpose and can serve as a clearinghouse for national media outlets—particularly in advance of an election campaign season.

Federal judges sitting in marble temples who are making public policy—rather than elected officials—should not get a pass either. You, along with the media, need to pin down legislators on whether or not they favor Constitutional reforms to limit the powers of judges that our Founding Fathers never intended them to exercise.

These reforms could include doing away with lifetime appointments by instituting term limits, revising the judicial confirmation process to include not only nominated judges but also a review of the conduct of sitting judges, and giving Congress the power to veto Supreme Court rulings. Although the passage of meaningful judicial reforms is unlikely in today's highly charged partisan environment, at least you will have a better idea of where candidates stand on judicial abuse of our Constitutional rights, or if they simply favor the status quo.

Once elected, politicians who *professed* to support property rights reforms during their campaigns need to be monitored. This oversight will send a message to ambitious politicians to think twice before they cast a vote in favor of seizing property through eminent domain for private gain or extorting money or property from owners seeking approvals to utilize their land.

Making our elected and appointed officials more accountable through the ballot box and monitoring their conduct are essential steps toward preserving our private property rights—rights our forefathers deemed sacred.

We, the People, endowed by our Creator with certain unalienable rights, must protect our property from government pirates! Let's get started.

APPENDIX: Legislative Reform Strategies

EMINENT DOMAIN

- Congress can pass legislation to make sure that federal funds are not used in eminent domain cases in which the objective is to generate more tax revenues.
- States can enact moratoria on the use of eminent domain to give citizens' groups time to lobby legislators and governors.
- State constitutions can be amended to prohibit the taking of property for private use.
- Local ordinance definitions of blight can be enacted to ensure that only property that is truly dilapidated, unsanitary, or hazardous to public health can be condemned.
- Citizens can press for new laws or regulations that would prohibit a city, county, or redevelopment agency from seizing property for the sake of generating more tax revenues.
- Cities and counties can pass ordinances prohibiting themselves from condemning private property and transferring it to private parties—except for utility projects.

NO-COMPENSATION AND PAY-TO-PLAY TAKINGS (ALSO KNOWN AS "REGULATORY TAKINGS")

- A Presidential Executive Order should be executed—modeled on one signed by Ronald Reagan—that directs

all federal agencies to "do no harm" to property rights
when issuing regulations.

- Property owners with constitutional claims should
  have direct access to federal courts and avoid having to
  go through expensive and time-consuming state court
  proceedings first.
- Legislation should be passed that is modeled after
  Oregon's *Measure 37* to force local and state governments to compensate owners in instances where their
  property has been devalued by land use regulations.
- Specific time frames should be instituted for permit approval or rejection, rather than leaving property owners
  in regulatory limbo for years.
- Loopholes that deprive property owners of *any* compensation unless the *entire* value of their property is
  reduced must be closed.
- The Clean Water Act should be amended to limit the
  jurisdiction of federal agencies over wetlands cases
  (except when toxic pollutants are involved) and to provide clear and reasonable guidelines that do not overly
  burden property owners.
- The Endangered Species Act should be amended to
  have recovery goals in place before species are listed
  and to provide both compensation for the loss of property value and incentives for landowners to conserve
  habitat.

ACKNOWLEDGMENTS

I **wish to** express my sincerest appreciation to my friend Sean Hannity. This book would not have been possible without his encouragement and unfailing support at each step of the way. More important even than our professional association, however, has been the close bond that has been forged between our families, including all the troops from Long Island. This I value above all else.

I owe a special thanks to Neal "The Talkmaster" Boortz. Neal's enthusiasm for *Government Pirates* helped spur me on to make it the very best I could. Neal is truly a great writer. In addition to fulfilling his role as America's "High Priest of the Church of the Painful Truth," Neal's ability to write about a dry subject and make it entertaining—as he did with the tax code in his *FairTax* books—reveals his true genius. I am proud to call him a close friend.

Barbara Malone at the Queensbury Press served as my pre-submission manuscript editor. Barbara played a major role in shaping and reshaping the book before it was shipped off to the Big Apple. Her thoughtful comments and criticism, along with those of her husband, Bob Malone, are greatly appreciated. Thanks to Isabelle Hart for a final proofing of the manuscript.

My attorney, David Limbaugh, took time from his busy legal practice and his publishing work as a bestselling author and nationally syndicated columnist to help guide me into the hands of HarperCollins. I am indebted to him.

From our first meeting, Cal Morgan, my editor at Harper-Collins, has been a steadfast advocate of the book. It has been a pleasure working with him, Brittany Hamblin, Jennifer Hart, Alberto Rojas, Nicole Reardon, and the other dedicated people in my HarperCollins family. I hope this is the first of many projects we do together.

Dick Corace, my brother, and my fellow partners at Signature Communities—Jerry Griffin, Keith Sharpe, and Glenn Griffin—deserve special recognition. Many of the government permitting nightmares that we have been through together over the years helped to serve as inspiration for this book.

Signature Communities is most fortunate to have a great friend and partner in Anthony Soave, one of America's great businessmen. Tony and the executives at Soave Enterprises continue to play an integral role in the success of our developments.

Although my parents, Jim and Jean Corace, have passed, there is not a day that goes by that I do not think about them. Dad, I miss those days of fishing the waters of Sanibel, Captiva, and Boca Grande and your sage advice—whether I asked for it or not. Mom, you would be glad to know that Ammi, the love of my life, shares your values and unselfish commitment to family. You would also be pleased that we have lived by your words of wisdom: "The most important thing for parents to do for their children is to love each another."

To my children, Natalie, Brandon, and Erik—each of you is unique and special in your own way. You know you can always count on your mom and me to be there for you and yours.

My most heartfelt thanks go to my wife, Ammi. Without your unconditional love and support, this book would not have been possible. Although our recent change to "empty nester" status has had its challenges, I believe that the best and most enriching chapters of our lives together lie ahead. I love you.

SPECIAL ACKNOWLEDGEMENTS

I wish to express my sincere gratitude to the people across America who contributed to my research and especially to those property owners whose stories of abuse cried out to be told:

- Margaret Cooper, attorney at Jones, Foster, Johnston & Stubbs, served as a major inspiration for the book. I am grateful for her valuable contributions to the Yardarm story and with the legal research on landmark cases.
- Jim Stephanis holds a special place in my heart. He displayed tremendous courage in permitting me to open up old wounds. His remarkable tale is one of the most egregious government abuse stories in America.
- Maria Hill and her son Jonathan deserve my thanks for sharing their family's story and shedding light on the tactics misguided preservationists use to steal property. The ordeal they suffered should not happen in America.
- Frank Kottschade knew he had an uphill battle when he took the City of Rochester, Minnesota, to court. Although he was seeking justice for the taking of his property through overregulation, Frank wanted to expose the inequalities of the ripeness doctrine. All builders, both large and small, owe him a debt of gratitude.
- Ocie Mills's fight to defend his constitutional rights landed him in prison. His battle shed a national light

on how government agencies abuse their power when it comes to wetlands regulation under the Clean Water Act. Ocie, you are a true patriot.

- The National Association of Home Builders deserves special recognition for its fine work. NAHB continues to serve as the voice for our nation's homebuilders. In addition to the many services it provides its 235,000 members, NAHB has vigorously defended the property rights of builders and has tirelessly lobbied for reforms. I am proud to be a member and stand ready to protect the interests of this vital industry.

- Attorney Jim Burling with the Pacific Legal Foundation is one of our country's most eminent legal experts on property rights. Jim opened many doors for me in the property rights community, and I am in his debt.

- Appellate lawyers Michael Berger and Gideon Kanner with Manatt, Phelps & Phillips have argued some of the most controversial property rights cases before the highest courts in the land. They have elevated legal brief writing to an art form. Thanks, guys, for helping this non-lawyer boil down complex legal concepts.

- Several not-for-profit property rights groups throughout the country deserve special recognition for their tireless efforts to curb property rights abuse. Among them are: Institute for Justice (and its related organization Castle Coalition), Pacific Legal Foundation, National Center for Public Policy Research, Cato Institute, Heritage Foundation, American Land Foundation, Liberty Matters, Mountain States Legal Foundation, Property Rights Foundation of America, and Coalition for Property Rights. I strongly encourage all readers to visit GovernmentPirates.com or DonCorace.com and click on "Links" to learn about these dedicated organizations. They depend solely on tax-deductible contributions. Please donate generously so that they can con-

tinue to represent victims of government abuse in court
and press for meaningful property rights reforms.

- The National Center for Public Policy Research in
  Washington, D.C., was the first not-for-profit organiza-
  tion I encountered when I began to research this book.
  David Ridenour, Peyton Knight, and Ryan Balis were of
  invaluable assistance and provided me with an "inside-
  the-beltway" perspective on the issues. The Center's
  *Shattered Dreams: One Hundred Stories of Government
  Abuse Stories* is an excellent compilation of stories that
  will make you ask yourself, "How can this happen in
  America?"

- Robert J. Smith is one of the most passionate and
  knowledgeable people in the property rights community. R.J. has an encyclopedic mind when it comes to
  property rights in general but particularly with issues
  related to endangered species.

- Carol LaGrasse of Property Rights Foundation of
  America is another devoted property rights advocate.
  I learned more in one-hour phone conversations with
  her than I could in several days of research. Keep up
  the good work, Carol.

- Dan Byfield of the American Land Foundation and Liberty Matters helped give me an "outside-the-beltway"
  perspective on how the rights of our nation's ranchers
  and farmers are abused. Dan helped with the Austin
  cave bug story—an example of how the courts have
  perverted the commerce clause in the U.S. Constitution.

- Attorney and author William Perry Pendley is another
  stalwart in the property rights arena. He heads up
  Mountain States Legal Foundation and has served as
  a champion for landowners in the western states for
  years. His work on the John Schuler case and many
  others must be commended. His latest book, *Warriors*

for the West, is required reading for anyone interested in the defense of property rights.

To all of you who are engaged in this ongoing battle, I hope that this work will serve the cause of justice. Thank you.

NOTES

INTRODUCTION: PROPERTY FOR THE TAKING

luxury hotel and upscale condos: Matt Apuzzo, "Supreme Court Rules Cities May Seize Homes," Associated Press, June 24, 2005.

A Sad Day for Property Rights: "The *New York Times* Loves Eminent Domain: Elite newspapers and liberal activists embrace the Kelo decision at their long-term peril," *Reason Online*, October 2005.

Souter's home: Norma Love, "States have a long history of taking land for economic reasons," Associated Press, September 5, 2005.

seizure of private property: "The Polls Are In: Americans Overwhelmingly Oppose Use of Eminent Domain for Private Gain," CastleCoalition.org.

seizures for private development: "The *New York Times* Loves Eminent Domain: Elite newspapers and liberal activists embrace the Kelo decision at their long-term peril," *Reason Online*, October 2005.

exclusive condo tower: Dana Berliner, Senior Attorney, Institute for Justice, CastleCoalition.org (source for all bullet points).

a bill to reach the Senate: "One Last Chance for Federal Eminent Domain Reform!" CastleCoalition.org, e-mail alert, September 29, 2006.

A restaurant owner: John Dorschner, "Hung from the Yardarm," *Tropic* magazine, *Miami Herald*, June 9, 1996.

a family's dairy farm: "The Squashing of Landowners' Property Rights," LandGrab@ma.unltranet.com.

A 70-year-old man: *United States v. Rapanos*, U.S. Supreme Court case commentary, PacificLegal.org.

In the Florida panhandle: "An Interview with Ocie Mills," *Liberty Matters Journal*, Fall 1999.

A developer in Austin: Dan Byfield, "Bugs Rule," *Cornerstone Newsletter,* Volume 9, Issue 3, July 2002.

an endangered species of fly: Ike C. Sugg, Competitive Enterprise Institute, "Lord of the Flies: the United States government is forcing landowners to spend millions of dollars to protect an endangered bug—Delhi Sands flower-loving fly," *National Review,* May 5, 1997.

1—THE ALMIGHTY JUDGES

the Supreme Court and federal courts: Alexander Hamilton, *The Federalist Papers,* Number 78.

maintaining the independent judiciary: Sandra Day O'Connor, "The Threat to Judicial Independence," Editorial, *Wall Street Journal,* September 27, 2006.

ultimate arbiter: Letter from Thomas Jefferson to W. H. Torrance, 1815 ME 14:303, *2000 Street Law, Inc.* and *Supreme Court Historical Society.*

the federal judiciary could review: Mark Levin, *The Men in Black: How the Supreme Court Is Destroying America* (Regnery Publishing, 2005), pp. 24, 25, 30, 31.

the center of controversy: "Implied Powers," u-s-history.com.

judicial tyranny: Mark Levin, *The Men in Black* (Regnery Publishing, 2005), p. 22.

The Super-Legislature: Mark Levin, *The Men in Black* (Regnery Publishing, 2005), pp. 195- 203.

2—THE DESPOTIC POWER

despotic power: James W. Ely, "Can the 'Despotic Power' Be Tamed?" Probate and Property, *American Bar Association,* December 2003.

The court concluded: Thomson/West, *Berman v. Parker* [Online] Available. Westlaw.

The decision ushered in: Thomson/West, *Poletown v. City of Detroit* [Online] Available. Westlaw.

the transfers were made: Gideon Kanner, "The Public Use Clause: Constitutional Mandate or Hortatory Fluff?," *Pepperdine Law Review,* p. 355.

ill-defined notion: "Landmark Eminent Domain Abuse Decision," *County of Wayne v. Hathcock,* CastleCoalition.org, Web Release, July 31, 2004.

when the government takes property: Supreme Court Justice
Clarence Thomas, Dissenting Opinion, *Kelo v. New London,
Connecticut*, June 23, 2005.

opened the floodgates: Dana Berliner, "Opening of the Floodgates: Eminent Domain in a Post-Kelo World," Castle Coalition report, June 2006.

10,000 examples: "One Year After Kelo: New Reports Document
Skyrocketing Eminent Domain Abuse, Chronicle the Legislative
Response to Kelo, and Expose the Myths and Development Fail-
ures," CastleCoalition.org, Web Release, June 20, 2006.

simply reaffirmed years of precedent: "The Kelo Decision: In-
vestigating Takings of Homes and Other Private Property," testimony before the Committee on the Judiciary United States
Senate, September 20, 2005.

90 percent of Americans polled: "The Polls Are In: Americans
Overwhelmingly Oppose Eminent Domain for Private Gain,"
CastleCoalition.org.

3—REDEVELOPMENT: IS IT A SCAM?

like playing Monopoly with real money: Steven Greenhut,
*Abuse of Power: How the Government Misuses Eminent Do-
main* (Seven Locks Press, 2004).

These redevelopment projects: Jerry Andrews, "David Versus Goliath in Humboldt County," *Downey Eagle*, September 17, 1999.

This corporate welfare: "Redevelopment: The Unknown
Government," *Municipal Officials for Redevelopment Reform*,
Second Edition, August 1998.

the city selling bonds: "A Primer on Tax Increment Financing
in Pittsburgh: Who Pays in TIFs," Neighborhood Capital Budget Group, *Allegheny Institute*, Report #99-06, June 1999.

many redevelopment projects fail: Jerry Andrews, "Redevelopment shortchanges schools," *Downey Eagle*, July 30, 1999.

volunteers make the repairs: "Think Successful Redevelopment Requires Eminent Domain? Think Again," CastleCoali-
tion.org, November 16, 2005.

This paved the way: Mark Brnovich, *Kelo, et al. v. City of New
London*, Amicus Brief, *Goldwater Institute*, p.6.

The project has spurred: "Fall Creek Receives Another National Reward," Office of Bart Peterson, Mayor of Indianapolis,
Testimony before U.S. House of Representatives, Committee
of the Judiciary, September 22, 2005.

4—A LITTLE PINK COTTAGE BECOMES A NATIONAL SYMBOL

one of the most vilified: "Battle Over Property Rights Goes On, Despite Ruling," *Christian Science Monitor*, January 4, 2006.

serious declines in employment: Lucette Lagnado, "Why New London, Conn., Still Waits for Its Ship to Come In: Pfizer's Vision for Research-Center Area Remains Far from Realized in Bitter Town," *Wall Street Journal*, September 10, 2002.

During the two-hour meeting: "Stop Eminent Domain Abuse," Testimony of former New London, Conn., Mayor Lloyd Beachy, Natural Rights-Org.

In early 2000: Susette Kelo, "Real People Pay Dearly for NLDC Land Grab," *The Day*, February, 11, 2001.

Within a year after Susette had bought: "Eminent Domain Without Limits? U.S. Supreme Court Asked to Curb Nationwide Abuses," Institute for Justice, Litigation Backgrounder, www.ij.org., July 2004.

the state would chip in $75 million: "Why New London, Conn., Still Waits for Its Ship to Come In: Pfizer's Vision for Research-Center Area Remains Far from Realized in Bitter Town," *Wall Street Journal*, September 10, 2002.

One dollar!: "A Threat to Property Owners," Editorial, *Hartford Courant*, March 5, 2004.

a watering hole: "Eminent Domain Without Limits? U.S. Supreme Court Asked to Curb Nationwide Abuses," Institute for Justice, Litigation Backgrounder, www.ij.org., July 2004.

Governor Rowland: Matt Apuzzo and John Christoffersen, "Former Gov. Rowland gets a year in prison for graft," Associated Press, April 18, 2003.

The tremendous social costs: "A Threat to Property Owners," Editorial, *Hartford Courant*, March 5, 2004.

the Institute for Justice filed its appeal: Chip Mellor, "The 25 Best Friends of Property Rights, 25 Amicus Briefs Filed at Supreme Court in IJ's Kelo Case," Institute for Justice publications, www.ij.org.

the five other property owners agreed: Elaine Stoll City, "Votes to Proceed with Property Seizures," *The Day*, June 6, 2006.

to preserve the home: "Susette Kelo Lost Her Rights, She Lost Her Property, But She Has Saved Her Home," Institute for Justice, Web Release, June 30, 2006.

5—THOSE PESKY HOLDOUTS

rent out: "Eminent Domain Abuse in Long Branch, N.J.: City Seeks to Kick Out Middle Class Along the Oceanfront to Make Way for the Rich," Institute for Justice, Litigation Backgrounder, www.ij.org.

to revitalize certain areas: "Norwood, Ohio Homeowners & Small Businesses Battle City & Private Developer Over Eminent Domain Abuse," Institute for Justice, Litigation Backgrounder, www.ij.org.

an inheritance: Gary Greenberg, "The Condemned: In Ohio, and across the country, homeowners are battling cities and developers conspiring to seize their property," Mother Jones. com, Jan./Feb. 2005.

If the city authorizes this study: Greg Peath, "Whose property is it?: To some, eminent domain use adds up to misuse," *Cincinnati Post* Online Edition, April 25, 2003.

he must have felt a collective vibe: Scott Bullock, "The Norwood, Ohio, Eminent Domain Trial," Institute for Justice, IJ Publications: Liberty & Law, www.ij.org.

The Ohio Supreme Court has never held: "Ohio Judge Upholds Use of Eminent Domain in Nice Neighborhood: Too Many Homeowners for City's Liking," www.ij.org, Web Release, June 14, 2004.

They opted to leave: "New Filing Urges Ohio Supreme Court to Protect Home from Eminent Domain Abuse," www.ij.org, Web Release. February 11, 2005.

The legal costs: Steve Kemme, "Myth vs. Reality: The Norwood case shatters some commonly held myths about eminent domain," *Cincinnati Enquirer* Online, April 30, 2006.

F*** the Gambles!: Gary Greenberg, "The Condemned: In Ohio, and across the country, homeowners are battling cities and developers conspiring to seize their property," Mother Jones.com, Jan./Feb. 2005.

inside a circle with a red line: Steve Kemme and Gregory Korte, "The Untold Story: The government wanted their homes and businesses for offices and shops. Six owners said no. And the dissension began." *Cincinnati Enquirer*, April 30, 2006.

the city spray-painted: "X Marks the Stop of Eminent Domain Abuse," Web Release, www.ij.org, June 20, 2005.

Our home is ours again!: "Ohio Supreme Court Rules Unanimously to Protect Property from Eminent Domain Abuse, Legislation Still Needed to Stop Rampant Abuse of Eminent Domain in Ohio," Institute for Justice Press Release, www. ij.org, July 26, 2006.

I can't wait to see my old place: Steve Kemme, "I am thrilled to own my house again," *Cincinnati Enquirer*, July 27, 2006.

This house was entrusted: "Video: Norwood homecoming," *Cincinnati Enquirer*, July 27, 2006.

6—MURKY WATERS IN RIVIERA BEACH

The people who live on the water: David Cauchon, "Pushing the limits of 'public use,'" *USA Today*, April 31, 2004.

The term "negro removal": "Clarence Thomas Cites 'Negro Removal' in Eminent Domain Case," *Cincinnati Black Blog*, June 25, 2005.

One-quarter of Riviera Beach's residents: "Florida: Riviera Beach Considers Eminent Domain Project," *Realtor Magazine Online*, April 28, 2006.

80 homes in the community: Pat Beall, *Palm Beach Post* Staff Writer, "Riviera Beach Eminent Domain Case Draws National Spotlight," *The eco.logic Powerhouse*, January 2006

only 342 actual homeowners: "Riviera Beach plan envisions a boatload of jobs," *The South Florida Business Journal*, October 3, 2005.

The green cottage: Pat Beall, *Palm Beach Post* Staff Writer, "Riviera Beach Eminent Domain Case Draws National Spotlight," *The eco.logic Powerhouse*, January 2006.

mostly black city council: Larry Keller, "Riviera downplays need to take houses for riverfront development," PalmBeach Post.com, December 18, 2005.

a resort developer: "Riviera Beach plan envisions a boatload of jobs," *The South Florida Business Journal*, October 3, 2005.

so-called blighted areas: "Litigation Backgrounder: City Officials Try Illegal Land Grab," Pacific Legal Foundation, PacificLegal.org.

The city's backroom development deal: "Injunction Sought to Block Riviera Beach's Illegal Land Grab—Backroom Development Deal Called 'Unlawful,'" Pacific Legal Foundation, PacificLegal.org, Press Release, June 12, 2006.

beauty salon business: "Riviera Beach, Florida Eminent Domain, *Wells v. Riviera Beach* Institute for Justice Cases," www.ij.org.

It was also reported: William Cooper, "Riviera consultants busting budget," PalmBeachPost.com, June 11, 2006.

Viking Inlet was forced to rethink: Brian Skoloff, "Developers may sue city, state in eminent domain case," Associated Press, October, 19, 2006.

the city council voted 5 to 0: Brian Skoloff, "Eminent domain project in limbo," Associated Press, November 18, 2006.

The dreamer was killed: William Cooper Jr., "Riviera Beach swears in mayor," PalmBeachPost.com, April 5, 2007.

It is not immediately clear: Brian Skoloff, "Lawsuits dropped after Riviera Beach promises not to use eminent domain," Associated Press, May 11, 2007.

7—STORY SNAPSHOTS

"The Donald" Trumped

Trump Plaza Hotel & Casino: Dana Berliner, "Property taking favors big guy over little guy," Laura Schoellkopf, *USA Today*, May 4, 1998.

The owners challenged: "Is Negotiation the Trump Card?" www.CastleCoalition.org, February 2, 2006.

Five Strikes and You're Out

Developer Moshe Tal submitted: "Government Takes Developer's Land for One Percent of Its Value, Sells It to Rival Developer," *Shattered Dreams: 100 Stories of Abuse*, National Center for Public Policy (4th Edition), 2003.

Tal then filed charges: "Antitrust: Racketeering Charges Properly Dismissed; Defendants Immune from Suit," Lexis Nexis Real Estate Report, Volume 1, Issue #3, July 2006.

99 Cents Is Not Enough

800-pound gorilla: Michael Berger, Attorney-at-Law, excerpts legal brief, *99 Cents Stores v. City of Lancaster.*

the naked transfer: James Ponnuru, "This Land Is Costco's Land," *National Review Online*, February 18, 2003.

Putting the Brakes on Eminent Domain Abuse

So when a local businessman: Sam Staley, "Wreaking Property Rights," ReasonOnline.com, February 2003.

Randy Bailey continues: "Mesa, Arizona, *City of Mesa v. Bailey*, Putting the Brakes on Eminent Domain Abuse," www.ij.org.

The New York Times: Corporate Welfare Recipient

We believe there could: Matt Welch, "Why the *New York Times* Loves Eminent Domain," ReasonOnline.com. October 2005.

ninety-nine-year lease: Jeff McKay, "Government, *NY Times* Join Forces to Evict Business Owners," CNSNEWS.com, October18, 2002.

Adding up all the tax breaks: Posted by Norman Oder, "More Coverage of the Times Tower eminent domain battle: from the *NY Sun*," TimesRatnerReport.blogspot.com.

Behind the Stadium Land Grab

twenty-four of the properties: Dana Berliner, Public Power, Private Gain, Institute for Justice, p.103.

nearly $270,000: "Sports Arena Hat Trick Penalizes Property Owner," *Shattered Dreams: 100 Stories of Abuse,* National Center for Public Policy (5th Edition), 2007.

Not a Very Bright School District

the district decided to sell: "Blast from the Past in Cumberland County!" *The Commonwealth Iconoclast* blog, posted February 9, 2007.

$30,000 in mortgage payments: "Examples of Eminent Domain Abuse in Virginia," Virginia Property Rights Coalition, www.vaproperty rights.org.

8—CONTROLLING LAND LOCALLY

In addition to restricting land use: "Real estate Law—Zoning—General Zoning Law Questions FreeAdvise, What is zoning?" www.realestate-law.freeadvise.com.

land use categories: "Sample Zoning Definitions, A few examples from Planning Advisory Report 421," American Planning Association.

explosive migration: Miriam Wasserman, "Urban Sprawl," Regional Review, *Federal Reserve Bank of Boston*, Quarter 1, 2000.

to construct highways: Chet Boddy, "A Brief History of Urban Planning—Part 2" www.chetboddy.com.

One landmark Supreme Court case: Margaret L. Cooper, analysis of Supreme Court of the United States, *Donald W. Agins et ux, v. City Tiburon,* [Online] Westlaw, *Thomson/West,* 2005.

clogging the federal courts: Gregory Overstreet, "The Ripeness Doctrine of the Taking Clause: A survey of decisions showing just how far federal courts will go to avoid adjudicating land use cases," *Journal of Land Use & Environmental Law*, 1994.

the owner filed an action: Margaret L. Cooper, analysis of Supreme Court of the United States, *Williamson County Planning Commission v. Hamilton Bank of Johnson City* [Online] Westlaw, *Thomson/West*, 2005.

to seek compensation: Michael Berger and Gideon Kanner, "Shell Game! You Can't Get There from Here: Supreme Court Ripeness Jurisprudence in Takings Cases at Long Last Reached the Self-parody Stage," *The Urban Lawyer*, University of Missouri–Kansas City School of Law, Fall 2004, Vol. 36, Number 4.

the landowner was entitled: Margaret L. Cooper, analysis of Supreme Court of the United States, *First Evangelical Lutheran Church v. County of Los Angeles* [Online] Westlaw, *Thomson/West*, 2005.

700 single-family lots: Michael Berger, excerpts from legal brief, *Tahoe-Sierra Preservation Council v. Tahoe Regional Planning Agency*, 2002.

John Roberts: Douglas T. Kendall, "What Makes Roberts Different," *Washington Post*, July 24, 2005.

9—CITY OF PIRATES

One of the greatest things: John Dorschner, "Hung from the Yardarm," *Tropic* magazine, *Miami Herald*, June 9, 1996.

a 120-day building moratorium: Earl Faircloth, Legal Brief (Summary of Testimony), attorney representing Yardarm.

To add insult: Personal interview with Jim Stephanis.

condo residents would not give up: John Dorschner, "Hung from the Yardarm," *Tropic* magazine, *Miami Herald*, June 9, 1996.

In a scathing opinion: John Dorschner, "Hung from the Yardarm," *Tropic* magazine, *Miami Herald*, June 9, 1996.

The jury's verdict: Ross Shulmister, "The Yardarm Saga, A complete history and analysis," *Ledger* (Pompano Beach), March 4, 1999.

Yardarm then made the last of several offers: Margaret Cooper, various letters, legal briefs, and court documents, Jones, Foster, Johnston, & Stubbs, West Palm Beach, Florida.

10—BUZZARDS CIRCLING OVER BUZZARDS BAY

Nestled at the head of the Slocum River: The Towns and Villages—Coastal Villages of Rhode Island and Massachusetts, www.coastalvillages.com.

The community has more than sixty miles: "Dartmouth, Massachusetts: A Brief History," www.town.dartmouth.ma.us/welcome.htm.

The family bought property themselves: Jonathan Hill, "The Squashing of Landowner's Property Rights," users.rcn.com/landgrab/summary.html.

A former town planner: Carol LaGrasse, "No Mercy: Preservation Group Straps Massachusetts Farm Couple," Property Rights Foundation of America, Inc., reprinted from *New York Property Rights Clearinghouse*, Vol.4, No. 2, Spring-Summer 1997.

Maria was aware of a state program: Interviews with Jonathan Hill and Maria Hill and court documents and notes provided by same.

unless the family sold FORM the property: Interviews with Jonathan Hill and Maria Hill and court documents and notes provided by same.

The Hills agreed that Huber: Interviews with Jonathan Hill and Maria Hill and court documents and notes provided by same.

According to their newsletter: "Saving Slocum's River in Dartmouth," *The Trustees of Reservations*, Quarterly Newsletter, Special Places Volume 6, No. 4, Fall 1998 & "Grassroots Support Spurs on Slocum's River Protection Campaign," *The Trustees of Reservations*, Special Places, Fall 1999.

11—HIGHWAY ROBBERY

He was born and raised: Testimony of Franklin P. Kottschade on Behalf of the National Association of Homebuilders in support of H.R. 4772, Private Property Rights Implementation Act of 2005, June 8, 2006.

His request was denied: Testimony of Franklin P. Kottschade on behalf of the National Association of Homebuilders in support of H.R. 4772, Private Property Rights Implementation Act of 2005, June 8, 2006.

he felt forced to sue the city: Testimony of Franklin P. Kottschade on behalf of the National Association of Homebuilders in support of H.R. 4772, Private Property Rights Implementation Act of 2005, June 8, 2006.

Jeffrey Pieters, "Developer awarded $1.5 million for lost drive-ways," *Post-Bulletin* (Rochester, MN), April 17, 2007.

The measure passed: H.R. 4128, 109th Congress, Private Property Rights Protection Act of 2005, GovTrack.us.

12—STORY SNAPSHOTS

Pay Ransom or Else

$1 million to renovate: Brian Watson, "The San Remo Hotel: Regulating the American Dream", Northwestern University—Medill School of Journalism, Medill News Service, March 2005.

During the oral argument: Mike McKee, "Dispute a Burr to Brown," Cal Law, California's Legal News Source, www.law.com, December 7, 2001.

A Twenty-Year Property Freeze

the proposed construction of new condominiums: Michael Berger, excerpts from legal brief, *Tahoe-Sierra Preservation Council v. Tahoe Regional Planning Agency*, 2002.

On appeal in 2002: "Court Declines to require payment for temporary taking of property," *Tahoe-Sierra Preservation Council v. Tahoe Regional Planning Agency*, Litigation Update, Washington Legal Foundation, www.wlf.org.

Because of the implications: Lake Tahoe (IJ Amicus), *"Tahoe-Sierra Preservation Council v. Tahoe Regional Planning Agency*, IJ Helps Challenge Unending 'Temporary Takings' of Property," Institute for Justice, www.ij.org.

It's Okay to Pray, But Not Too Often

In October 1998: "Couple's Home Bible Study Banned by Zoning Ordinance," National Directory of Environmental & Regulatory Victims, National Center for Public Policy, 2000.

Monday Night Football: Rev. Jerry Falwell, "1999: Year of Persecution," WorldNetDaily.com, posted January 8, 2000.

paid $30,000 for all court costs: Matt Kantz, "Denver couple wins right to home prayer meetings," *National Catholic Reporter*, Dec. 31, 1999.

Homeowners with a View, Beware

620-foot Multnomah Falls: Sam Howe Verhovek, "Dream House with a Scenic View Is Environmentalist's Nightmare," *New York Times*, April 24, 1999.

bi-state government agency: "Home at Last for Brian and Jody Bea!," Pacific Legal Foundation bulletin, www.pacificlegal.org.

Not in Their Backyard

For the past thirty years: "California Coastal Commission Claims Picnic Tables Will Deter Visitors to a Public Beach," *Shattered Dreams: One Hundred Stories of Abuse*, National Center for Public Policy (5th Edition).

Even if it weren't our property: Brian Martinez, "Not in Their Back Yard? State Asks O.C. Couple to Get Rid of Installations on Their Slice of Beach," *Orange County Register*, May 19, 2004.

Pay Me Now, or Pay Me Later

approved for 126 homes: Prepared Statement of Richard Reahard before the House Judiciary Committee on the Constitution Subcommittee—H.R. 2372, The Private Property Rights Implementation Act of 1999, *Federal News Service*, September 16, 1999.

one single-family residence: *Reahard v. Lee County*, United States Court of Appeals for the 11th Circuit, opinion, decided August 14, 1992.

Kids at Play, Keep It That Way

In early 1997: "$58,000 Spent Fighting Over Treehouse," *Shattered Dreams:100 Stories of Abuse*, National Center for Public Policy (5th Edition).

76 percent of registered voters: "Clinton Treehouse Proves 'You can fight city hall,'" Saveourtreehouse.com.

(Note: The title of this story was taken from a sign that had been placed on the tree house.)

13—REGULATING PUDDLES AND PONDS

one of the agencies authorized: Travis Hale, "Takings and Wetlands: Property Rights, Public Stewardship, and Compensation," December 3, 2001.

100,000 acres per year: Beth Baker, "Washington Watch: Government Regulation of Wetlands Is Under Siege by All Sides," American Institute of Biological Sciences, November 1999.

swath of marshes: Donald Smith, "Reclaiming the Florida Everglades," *National Geographic News*, December 2000.

$14 billion program: "The Lost Coast," Louisiana's Wetlands

Map at *National Geographic Magazine*, magma.nationalgeo-
graphic.com.

it created a regulatory plan: Todd H. Votteler, University
of Texas and Thomas A. Muir, National Biological Service,
"National Water Summary on Wetland Resources, Wetland
Management and Research: Wetland Protection Legislation,"
U.S. Geological Survey Water Supply Paper 2425.

The Corps of Engineers sought: *United States v. Riverside
Homes, Inc.*, U.S. Supreme Court, opinion, decided December
4, 1985.

a city attempted to exert its authority: *Dolan v. City of Tigard*,
U.S. Supreme Court, majority opinion, decided June 24, 1994.

the ponds ranged in size: Robert W. Thomson, Buchanan In-
gersoll, P.C., "Migratory Bird Rule Upheld," 2000.

The appeals court ruled: *Solid Waste Management of Northern
Cook County v. U.S. Army Corps of Engineers*, United States
Court of Appeals, 7th District, opinion, decided January 9,
2001.

The high court ruled: *Solid Waste Management of Northern
Cook County v. U.S. Army Corps of Engineers*, U.S. Supreme
Court, majority opinion, decided January 9, 2001.

One reason these disputes: Travis Hale, "Takings and Wetlands: Property Rights, Public Stewardship, and Compensation," December 3, 2001.

14—A WETLANDS DESPERADO

He worked as a carpenter: Ocie Mills. . .In His Own Words,
ConstitutionalRightsNews.com.

redneck: Phone interview with Ocie Mills and court documents.

heart attack: Chris Lavin, "At War over Wetlands/Father and
Son Imprisoned After Fight with Government," *St. Petersburg
Times*, November 10, 1992.

He bought two lots: Henry Lamb, WorldDaily.net, posted September 22, 1999.

Another juror: "Freedom Is Not Free: An Interview with Ocie
Mills," *Liberty Matters Journal*, Fall 1999.

In January 2007: Phone interview with Ron Johnson, attorney
with Kinsey, Troxel, Johnson & Walborsky, Pensacola, Florida.

The order demanded: Wetland Conservation Act, Cease and
Desist Order #864872, Minnesota, Department of Natural
Resources.

15—THOSE DAM BUREAUCRATS!

The following is the letter: Barbara and David P. Mikkelson, "Damned Beavers!" www.snopes.com.

16—A WETLANDS NIGHTMARE

If even one molecule: Rapanos video, Pacific Legal Foundation, www.plf.com.

Greek immigrant: Nolan Finley, "Feds destroy lives and property rights," *Detroit News,* May 9, 2004.

forty-eight to fifty-eight acres of wetlands: *United States v. Rapanos,* United States Appeals Court for the Sixth District, opinion, decided and filed July 26, 2004.

Harding had visited: Cheryl Wade, "Former DEQ director: Michigan in tough straits over wetland laws," *Midland Daily News,* May 11, 2006.

Other experts confirmed: "*Rapanos v. United States:* Background, Blog, and Briefs," Pacific Legal Foundation, PacificLegal.org.

So here we have a person: David Stirling, "How 'wetlands' bureaucrats crush private-property rights," WorldNetDaily.com, August 17, 2004.

He cited a case: United States District Court of Eastern Michigan Southern Division, Judge Lawrence P. Zatkoff, opinion and order, March 15, 2005.

Government prosecutors sought: "*Rapanos v. United States:* Background, Blog, and Briefs," Pacific Legal Foundation, PacificLegal.org.

The Army Corps denied a permit: Patrick J. Wright and Russ J. Harding, "Michigan Landowners to Be Heard at U.S. Supreme Court," Mackinac Center for Public Policy, January 3, 2006.

Attorney General Mike Cox supported: "States, Enviros, Side with Feds on Supreme Court Clean Water Cases," *Environmental News Service,* January 19, 2006.

The Corps has not drawn: *Rapanos and Carabell v. United States,* U.S Supreme Court transcript, Alderson Reporting Company, Washington, D.C., Feb.21, 2006.

Justice Kennedy issued his own opinion: "Clean Water Act Reach Limited: U.S. Supreme Court Overview," Bloomberg.com, June 19, 2006.

17—STORY SNAPSHOTS

Dumped on over a Dump

The day after Thanksgiving: "Man Serves Hard Time for Cleaning Dump, Environmentalists Complain Government Has Been Too Easy," *Shattered Dreams: One Hundred Stories of Abuse*, National Center for Public Policy (4th Edition), 2003.

One EPA official: Robert H. Wayland, EPA, testimony before the Committee on Government Reform, U.S. House of Representatives, October 6, 2000.

Pozsgai's fine was reduced: "Wetlands: Pozsgai Fine Cut 97 Percent," *Greenwire*, Natural Resources, January 29, 1992.

Squashing a Property Owner's Rights

the Roberges were offered: "Corps of Engineers Tries to Set Example—But Instead They Are Made an Example of," National Directory of Environmental & Regulatory Victims, National Center for Public Policy, 1998.

With the support of a Maryland-based property rights group: Richard Miniter, "Contract Out on Bureaucratic Property Abuses," *Washington Times*, December 11, 1994.

The only action taken: "A Victory for Private Property," *Washington Times* (editorial), December 15, 1994.

sensitivity training: John S. Day, "Army Corps Tries, Fails to 'Squash' Elderly Pair," *Bangor Daily News*, August 8, 1996.

A Permit . . . But Not Really

Helen and William Cooley owned: "Corps of Engineers Takes Use of Couple's Property, Then Ducks Responsibility in Court with Monopoly Money Permit," *Shattered Dreams: 100 Stories of Abuse*, National Center for Public Policy (4th Edition) 2003.

The Cooleys declined the offer: Liddell & Sapp LLP, "Court Final on Wetlands Taking," *Texas Environmental Compliance Update*, September 18, 2000.

the Cooleys were awarded: Arnall Golden & Gregory, "Federal Claims Court Awards Couple $2 Million for Corps' Taking of Property," *Georgia Environmental Law Letter*, Volume 12, Issue 2, September 18, 2000.

No Good Deed Goes Unpunished

a dedicated conservationist: "California Winemaker Penalized for Creating Wetlands," *National Directory of Environmental & Regulatory Victims*, National Center for Public Policy, 2000.

more than 10,000 waterfowl: R. J. Smith, "Private Conservation Spotlight: Viansa Winery Wetlands," Heartland Institute,
Environmental News, August 1999.

Sunk by a Mud Puddle
the Olsens purchased: "Mud Puddle Ruled Protected Wetland,
Ruining Elderly Couple's Retirement Plans," *National Directory of Environmental & Regulatory Victims*, National Center
for Public Policy, 2000; Walter H. Olsen, Sr., speech before
Property Rights Foundation of America, Third Annual N.Y.
Conference on Private Property Rights, 1998.

Bogged Down by the EPA
farmed cranberries since 1920s: "PLF Asks Appeals Court to Rein
in Wetlands Regulators' Campaign Against Cape Cod Cranberry
Farmers," Pacific Legal Foundation, www.pacificlegal.org.
"the stream": Peter Schworm, "Wetlands Ruling Will Be Re-
heard Supreme Court Decision Cited," *Boston Globe*, South,
November 22, 2006.

A Pond for the Greater Good
the couple purchased four acres: "Local Government Takes
96% of Couple's Land," *National Directory of Environmental &
Regulatory Victims*, National Center for Public Policy, 2000.

18—THE ENDANGERED SPECIES ACT GONE WILD!
we are facing: Excerpts from video clip, "The Endangered Species Act," Penn & Teller, *Bullshit!*, Showtime.
It's a complete fabrication: Excerpts from video clip, "The Endangered Species Act," Penn & Teller, *Bullshit!*, Showtime.
an estimated 227 animal and plant species: "The Assault of
the Endangered Species Act: What Academic, Government,
and Independent Sources Really Say About Our Endangered
Species Law," The National Environmental Trust, www.net.
org, 2005.
487 active nesting pairs: Federal Register Notice, Proposed
Rules to Delist the Bald Eagle, June 5, 2007.
widespread shooting: "Endangered Species: Back From The
Brink, the Eagle Is Back," BackFromThe Brink.org, Environmental Defense Network, 2005.
incubating parents: "What Are Endangered and Threatened Species," U.S. Environmental Protection Agency, www.epa.org.

It was then listed: "Cetaceans, Dolphins, and Porpoises (Overview)," National Ocean and Atmospheric Administration—Office of Protected Resources, www.nmfs.noaa.gov.

The foundation's research claims: Alexander F. Annett, "Research: Energy and Environment," The Heritage Foundation, Heritage.org/Research/EnergyandEnvironment.

A National Environmental Trust (NET) report: "The Assault of the Endangered Species Act: What Academic, Government, and Independent Sources Really Say About Our Endangered Species Law," The National Environmental Trust, www.net.org. 2005.

its foraging area: Eric D. Forsman, "Northern Spotted Owl," U.S Forest Service, Pacific NW Research Station, Biology. usgs.gov.

In a 1995 Supreme Court case: U.S. Supreme Court, *Babbitt v. Sweet Home Chapter of Communities for a Great Oregon*, Opinion, decided June 29, 1995.

The wolves were listed as an endangered species: U.S. Court of Appeals, 4th District Court, Opinion, *Gibbs v. Babbitt*, Opinion, decided June 6, 2000.

The court cited: U.S. District Court of Appeals, District of Columbia, *Rancho Viejo v. Norton*, Opinion, filed July 22, 2003.

Scientific evidence proved: U.S. Supreme Court, *GDF Realty Investments v. Norton*, U.S. Supreme Court, Opinion, September 2004.

Michael Bean: Laura E. Huggins, "A better way to protect endangered species," Hoover Institute, February 10, 2003.

These landowners just don't exist: Excerpts from video clip, "The Endangered Species Act," Penn & Teller, *Bullshit!*, Showtime.

19—HELD HOSTAGE BY A FLY

eight-to-eleven-mile radius: Final Recovery Plan for the Delhi Sands Flower-Loving Fly, U.S. Fish and Wildlife Service, Pacific Region, 1997.

The fly's total life span: Delhi Sands Flower-Loving Fly, Life History and Recovery Activities, *U.S. Fish and Wildlife Service*; "Endangered Fly Stalls Some California Projects," (Displaying Abstract), *New York Times* (National Desk), December 1, 2002.

eight flies buzzing: Ike C. Sugg, Competitive Enterprise Institute, "How Protecting a Fly Can Hurt the Sick," *Wall Street Journal*, February 11, 1997.

jokes about killing: Ike C. Sugg, "Lord of the Flies: the United States government is forcing landowners to spend millions of dollars to protect an endangered bug—Delhi sands flower-loving fly," *National Review*, May 5, 1997.

fly preserve: Ike C. Sugg, "Lord of the Flies: the United States government is forcing landowners to spend millions of dollars to protect an endangered bug—Delhi sands flower-loving fly, *National Review*, May 5, 1997.

The USFWS then found: Environmental Assessment on Incidental Take for the Delhi Sands Flower-Loving Fly, Department of Interior, August 3, 1995.

Living room sofas: Matthew Heller, "A Simple Case of Insecticide," *Los Angeles Times*, February 16, 2003.

Colton was not the only city: Sandra Stokley and Adriana Chavira, "Fly Habitat Part of Settlement," *Press-Enterprise*, August 18, 2000.

The Agua Mansa Enterprise Zone: The Profit Zone, Agua Mansa Enterprise Zone, www.agua-mansa.com.

Property owners with bank debt: Ike C. Sugg, "Lord of the Flies: the United States government is forcing landowners to spend millions of dollars to protect an endangered bug—Delhi sands flower-loving fly," *National Review*, May 5, 1997.

channels of interstate commerce: District Court of Appeals of the District of Columbia, *National Association of Home Builders v. Norton*, Majority Opinion, decided July 8, 2005.

alleviate traffic congestion: Congressional Hearing held in Fontana, California, on the Impact of the Endangered Species Act on the Inland Empire, Off-Road Business Association, orba.biz.com.September 10, 2004.

staunch opponents: Editorial Advisory, Xerces Society for Invertebrate Conservation, www.xerces.org, August 2004.

A year later: Jody Clarke, "ESA Reform Bill Passes House: A New Era for Property Rights Protection and Endangered Species," Competitive Enterprise Institute, September 29, 2005.

important ecological role: Shawnetta Grandberry, U.S. Fish and Wildlife Service, Carlsbad, California, Office and Chris Nagano, Supervisor, Fish and Wildlife Biologist, Albuquerque, New Mexico, "Protecting a Flower-Loving Fly," *Endangered Species Bulletin*, September 1998.

$175 million: Stephan Wall, "City Pushing to Protect Fly," *San Bernardino County Sun*, January 15, 2007.

20—THE CAVE BUGS ARE SAFE

I don't want to negate: Scott Gold, "Rare Bugs Arrest a Development," *Los Angeles Times*, March 28, 2005.

narrow entrance of a cave: Dan Byfield, "Bugs Rule," *Cornerstone Newsletter*, Volume 9, Issue 3, July 2002.

30,000-area preserve: "ALF Challenges Endangered Species Act," *Standing Ground* newsletter, American Land Foundation, July 2000.

$10 million: U.S. District Court for the Western District of Texas, Austin Division, Plaintiff's Original Complaint, *GDF Realty Investments v. Babbitt*, March 6, 2003.

Fred Purcell attempted to clear brush: Scott Gold, "Rare Bugs Arrest a Development," *Los Angeles Times*, March 28, 2005.

HCPs protect the areas: Habitat Conservation Plans, Section 10 of the Endangered Species Act, U.S. Fish and Wildlife Service.

the USFWS refused to process: U.S. District Court for the Western District of Texas, Austin Division, Plaintiff's Original Complaint, *GDF Realty Investments v. Babbitt*, March 6, 2003.

Fred Purcell and his partners: Interview with Dan Byfield, American Land Foundation, June 18, 2007.

The U.S. District Court ruled: Dan Byfield, "ESA Challenge Moves to Fifth Circuit," Press Release, www.LibertyMatters. org, October 5, 2001.

Lawyers for the USFWS: U.S. Court of Appeals for the Fifth District, *GDF Realty v. Norton*, Majority Opinion, March 26, 2003.

"Hail Mary": Anton Caputo, "Sides Await Word on Cave Bugs Case: Supreme Court Could Decide Today Whether to Put It on the Docket," *San Antonio Express News*, January 24, 2005.

hundreds of species: "More Commerce Clause Craziness," *Boots and Sabers* blog, January 24, 2005.

essentially unlimited authority: "WLF Urges Supreme Court to Review Important Endangered Species Case (*GDF Investments v. Norton*)," Washington Legal Foundation, Press Release, September 8, 2004.

seventeen-year battle: Scott Gold, "Supreme Court Swats Down Texans' Effort to Weaken Species Protection Law," *Los Angeles Times*, June 14, 2005.

Despite what the Constitution says: Scott Gold, "Rare Bugs Arrest a Development," *Los Angeles Times*, March 28, 2005.

21—A GRIZZLY LESSON TO LEARN

On a snowy September: William Perry Pendley, "Doing Everything by the Book," AmericanPolicy.org; *Shuler v. Babbitt*, U.S. District Court for the District of Montana, *Shuler v. Babbitt*, decided March 17, 1998; William Perry Pendley, *Warriors for the West: Fighting Bureaucrats, Radical Groups, and Liberal Judges on America's Frontier*, Chapter 1, "Fighting the 'Pit Bull of Environmental Laws'" (Regnery Publishing, 2006).

The bear quickly rose: William Perry Pendley, *Warriors for the West: Fighting Bureaucrats, Radical Groups, and Liberal Judges on America's Frontier*, Chapter 1, "Fighting the 'Pit Bull of Environmental Laws'" (Regnery Publishing, 2006).

An adult male stands: Grizzly Bear (*Ursus arctos horribilis*), U.S. Fish and Wildlife Service, fws.gov/endangered/.

40 miles per hour: Congresswoman Helen Chenoweth, "The Dangers of the Proposal of the U.S. Fish and Wildlife Service to Introduce Grizzly Bears into Idaho," July 28, 1997.

The use of pepper spray: Anthony Acerrano, "The Traveling Hunter, Grizzly Defense—What's the best way to defend yourself if you run into a grizzly while hunting—or if a grizzly tries to run into you?," www.sportsafield.com.

After an eight-year battle: William Perry Pendley, *Warriors for the West: Fighting Bureaucrats, Radical Groups, and Liberal Judges on America's Frontier*, Chapter 1, "Fighting the 'Pit Bull of Environmental Laws'" (Regnery Publishing, 2006).

A few months later: William Perry Pendley, *Warriors for the West: Fighting Bureaucrats, Radical Groups, and Liberal Judges on America's Frontier*, Chapter 1, "Fighting the 'Pit Bull of Environmental Laws'" (Regnery Publishing, 2006).

22—STORY SNAPSHOTS

The Courts Don't Give a Hoot

a retired postal worker: James J. Kilpatrick, "Case of Spotted Owl Moves to High Court, *Augusta Chronicle* (Georgia), Editorial, October 10, 2002.

"Little Wiley": Department of Forestry Final Order, State of Oregon, Final Order, "In the Matter of the Denial of a Written Plan submitted by Marsha and Alvin Seiber," November 7, 2000.

It cited a case: "Taking Back Community Rights," *Community Rights Counsel* newsletter, June 2004.

Much of the timber: James J. Kilpatrick, "Case of Spotted Owl Moves to High Court," *Augusta Chronicle* (Georgia), Editorial, October 10, 2002.

A Different Kind of Beetlemania

Chesapeake Bay: "Maryland Hugs a Bug," *Washington Times*, Editorial, September 9, 1991.

a voracious hunter: Michelle Babione, "Bringing Tiger Beetles Together," *Endangered Species Bulletin*, January/February 2003.

Biologist Barry Knisley: Raymond McCaffrey, "In Calvert County, Efforts to Aid Endangered Species Leave Beachfront to Bugs," *Washington Post*, April 2, 2002.

A Secure Border Is for the Birds

2004 biological study: "California Agency Says a More Secure Border Is for the Birds," *Shattered Dreams: One Hundred Stories of Abuse*, National Center for Public Policy (5th Edition), 2007.

deaths of three Border Patrol agents: "Plan for U.S.-Mexico border fence rejected," CNN.com, February 19, 2004.

Stymied by Snails

Brandt Child's 450-acre property: Catherine Wang, "The Grass Is Looking Greener for Landowners," *BusinessWeek*, Top of the News; Number 3274, page 3.

one-inch, slimy mollusk: Jim Woolf, "Landowner, Feds Hope They Don't Have to Slug It Out over Endangered Snail," *Salt Lake City Tribune*, April 14, 1995.

shoot the geese: David Rothbard and Craig Rucker, "Troubled ESA endangers property owners, wildlife: Property Owners want regulatory ESA beast tamed," Committee for a Constructive Tomorrow, August 1, 1998.

$2.5 million: Rick Henderson, "Preservation Acts—environmental laws and the right of property: The Property-rights movement steps out of the shadows," *Reason*, October 1994.

The Mouse That Never Was

twenty-five miles away: "Possibly Non-Existent Mouse Shatters Family's Dreams," *Shattered Dreams: One Hundred Stories of Abuse*, National Center for Public Policy (5th Edition), 2007.

A curator: Mead Gruver, "Endangered Mouse Never Existed," Associated Press, June 11, 2004.

Snakes in the Grass

The project received local approval: Carol LaGrasse, "Wealthy Land Trusts and DEC Squeeze Fishkill Business Owner," *New York Property Rights Clearinghouse*, Vol. 4, No.2, Spring-Summer 1997.

Scenic Hudson: "Animal vs. Mineral: Mining is proposed for snake habitat," *New York Times*, Metropolitan Desk, October 21, 1998.

a species listed as "threatened": "Spitzer and DEC Win Court Order for Threatened Species: Unauthorized Fence at Dutchess County Stone Quarry Endangers Rare Timber Rattlers," Press Release, Office of New York State Attorney General, March 5, 1999.

The Bald Eagle: Delisted in Name Only

President Clinton: "Despite Bald Eagle Delisting, New Eagle Rule Burdens Land Use Too Much," Press Release, Pacific Legal Foundation, PacificLegal.org, June 4, 2007.

I can't even cut firewood: Peter Slevin, "Bald Eagle to Be Taken Off Endangered Species List," *Washington Post*, December 25, 2006.

330-foot radius: "Despite Bald Eagle Delisting, New Eagle Rule Burdens Land Use Too Much," Press Release, Pacific Legal Foundation, PacificLegal.org, June 4, 2007.

There is no greater tribute: "Comments about the Bald Eagle Being Delisted," Pacific Legal Foundation, PacificLegal.org.

23—A CALL TO ACTION

With public sentiment: Quote from Pacific Legal Foundation for *PLF Alerts*.

sentimental value: Timothy Sandefur, *Cornerstone of Liberty: Property Rights in the 21st Century* (Cato Institute, 2006), pp. 1, 12-14, 48.

The EPA's action: Timothy Sandefur, *Cornerstone of Liberty: Property Rights in the 21st Century* (Cato Institute, 2005), pp. 95, 102; Derek M. Johnson, *"Kelo v. New London:* A Legal Perspective," *The Connecticut Economy*, Winter 2006 issue; U.S. Supreme Court, *Ruckelhaus v. Monsanto Company*, Opinion, decided June 26, 1984.

courteous letters: Neal Boortz and John Linder, *The FairTax Book* (Regan Books/ Harper Collins, 2005), p. 178

These reforms: Mark Levin, *Men in Black: How the Supreme Court Is Destroying America* (Regnery Publishing, 2005), pp. 201, 202.

INDEX